T0092258

THE ALTRUISTIC
URGE

THE ALTRUISTIC URGE

WHY WE'RE DRIVEN TO HELP OTHERS

STEPHANIE D. PRESTON

Columbia University Press *New York*

Columbia University Press
Publishers Since 1893
New York   Chichester, West Sussex
cup.columbia.edu

Library of Congress Cataloging-in-Publication Data
Names: Preston, Stephanie D. (Stephanie Delphine), author.
Title: The altruistic urge : why we're driven to help others /
Stephanie D. Preston.
Description: New York : Columbia University Press, [2022] |
Includes bibliographical references and index.
Identifiers: LCCN 2021041849 (print) | LCCN 2021041850 (ebook) | ISBN
9780231204408 (hardback) | ISBN 9780231555524 (ebook) Subjects: LCSH:
Altruism. | Helping behavior.
Classification: LCC BF637.H4 P733 2022 (print) | LCC BF637.H4 (ebook) |
DDC158.3— dc23

Cover design: Noah Arlow
Cover image: Shutterstock

Thank you to Brent and my beloved girls

CONTENTS

PREFACE

People say that the wackiest news comes from Florida. In just a quick sampling of online collections of these bizarre stories, you might encounter headlines like these:

- Florida man steals a car, realizes a baby is in it, drops baby off safely, and makes his getaway.
- Florida man arrested for driving stolen vehicle while monkey clings to chest.
- Florida man breaks *into* jail to hang with friends.
- Florida man steals bees because he thought they were "abandoned."
- Pregnant woman rescues husband from shark attack in Florida.
- Florida man rescues puppy from jaws of alligator without dropping cigar.[1]

Besides showcasing the amusing, concerning, and even inspiring antics of Floridians, these actual stories share something else in common: they exhibit features of an evolved and natural urge to approach and care for those we are bonded to, care about, or perceive to be similar to a helpless infant—an urge that we sometimes even extend to adult strangers, pets, and wild animals.

This rescue instinct is often displayed in our retellings of human heroism, like the pregnant woman from Florida who saw a dorsal fin and her husband's blood in the water and dove in "without hesitation" to pull him to safety, or Wesley Autrey, who dived onto subway tracks in New York City to save a young man from an oncoming train after the man fell into the tracks after a seizure.[2] Such heroic rescues are even observed in other species, like the dog in Trinidad that saved his human companion from a house fire by barking and tugging at the man's pant leg until the man awoke. The same dog then died after running back into the burning home, perhaps to retrieve the pet parrot.[3]

These individuals seem heroic, if not a little crazy or even stupid. How could a species evolve a predisposition to help others in a way that could endanger their very life? Why does this tendency exist across species? What does this urge have to do with the empathy, helpful personalities, or deep thoughts that we usually associate with our unique human capacity to give? Conversely, if we are so recklessly driven to save others, how can we also turn such a blind eye toward others' suffering, all over the world?

I wrote this book to describe the nature of a specific form of altruism, which I call the altruistic urge, through an integrated theory of its evolution, psychology, and neural bases captured by the altruistic response model. Much ink has been spilled describing our general capacity for empathy, altruism, or even human morality writ large. This book is different because I decidedly do *not* try to explain our broad swath of human goodness. I simply make the argument that there exists a specific type of altruism that has persisted in our genome for quite a long time and exists across species, one that powerfully influences our motivation to help—even heroically. This particular form of altruism—the altruistic urge—has yet to be explored and, and such, deserves our dedicated time and attention to determine what it is, when it happens, and—as important—when it does not. This book thus explains the altruistic response model while combatting common concerns with such a proposal, such as: *I am not like a rat! I am not a caring person or mother! I do not feel urges to help! People are terrible, so your theory must be wrong! Do you think this explains all of altruism . . . because it doesn't! We help others because it was necessary to win wars!* And so on.

CATEGORIZING ALTRUISM

There are many kinds of altruism, but the fact that we don't have names for all of them causes confusion. I want to resolve this at the outset so that readers know what this book is (and is not) about. The problem is not so much with the science but with our impoverished semantics, which I wish were more precise, in the way that ornithologists create taxonomies of bird species with well-defined groups and names. My graduate school professor Eleanor Rosch explained that people naturally categorize things like birds into different levels of abstraction.[4] Most of us share the general concept of a "bird" as a small, winged animal, which we represent in our minds as the average of all birds we have encountered. For example, North Americans might think of a typical "bird" as a passerine or songbird, like a sparrow or cardinal. Birds from other orders within the avian class share a common ancestor with songbirds, which causes them to all be grouped into the same vertebrate class; however, some birds look more different from the typical bird than others, like flightless penguins and ostriches. Because of this, if you saw an ostrich walking down the street, you might exclaim, "Look! An ostrich!"

rather than "Look! A bird!" because even if ostriches are technically birds, you would recognize that the general term would
confuse your friend, who might look up into the trees rather than
at the street. People with bird expertise also use bird names that
are more specific than those used by nonexperts, such that a regular person who is not particularly interested in bird speciation
might point out the "pretty bird" to a friend on a walk, whereas
married people who watch birds from the kitchen table every
morning might refer to the "red songbird" or "Charlie," while
birders would whisper excitedly to each other about the wood
warbler or indigo bunting. There are different levels at which you
can specify what you mean when you refer to birds, and people
intuitively understand this and shift how they refer to them
depending on their knowledge, audience, and situation.

The concept of altruism is in many ways like the situation with
birds. There are lots of different kinds. Some forms of altruism
are more like what you would consider the typical type, whereas
others may be observed only in specific environments. When
most people hear the word "altruism," they might imagine a saint
who relinquishes worldly possessions to feed the poor, or a hero
who rescues a stranger from a burning building. A biologist may
think about the alarm calls of ground squirrels or the aid that
worker bees give to their queen. An economist may think about
how many dollars a student donates to a stranger in a laboratory
experiment. I might even consider the warm hug offered to a distressed friend. Even if you recognize each of these behaviors, you
personally might not apply the label "altruism" to all of them.
But unlike with the birds, even experts of altruism do not really
follow an agreed-upon taxonomy of types of altruism that you
could find in a textbook or memorize for a class. When scientists do try to make subdivisions of types of altruism, they
usually draw the lines between species or behaviors that look

dissimilar. Originally species were also divided up this way—putting birds together that looked similar to one another. But at this point, biologists also decide what counts as a bird from the evidence derived from fossils and wing morphology (even if penguins never come to mind when people think about birds).

Something like this needs to happen for altruism. We need to define our taxonomies based on evidence from things that include the way the behavior looks or its function, while also considering when they evolved and how they are mediated in the brain and body. For example, even if an ant can free another trapped ant in much the same way that a person can free the evil man's victim who has been roped to the train tracks, are the two the same? Do they rely upon the same neurophysiological mechanisms? Did they evolve from a common gene or set of genes for a similar purpose? Moreover, things that look dissimilar but are present in the same species or period of development perhaps should not be grouped together if they did not emerge at the same time, with the same mechanism. Thus, people assume that anything a human or great ape can do that a monkey or dog cannot do must represent a single emergent process that requires a large brain. But many of these lauded behaviors are also present in birds or rats, who possess very small brains indeed. Thus, just as you do not want to assume that pterodactyls, ravens, bats, and butterflies belong together and share a common ancestor because they all have wings and fly, we do not want to assume that forms of altruism go together unless we examine the evidence from across biology, psychology, and neuroscience.

This book is focused on one specific type of altruism, which my research reveals to be a natural kind in the taxonomy of acts of aid. The *altruistic urge*, as I call it, refers to events where any animal or person feels compelled to approach a vulnerable victim in immediate need of aid. This urge to respond appears to be evolutionarily ancient—like the wing of the bird that traces back to the therapod dinosaurs. The altruistic urge is different from the forms of altruism that are often described in the social sciences, which focus on a unique human capacity to consciously contemplate a decision to give. The altruistic urge does not make us special. It makes us more like other species, in fact. Sorry about that. However, if we understand this specific and powerful motivation, we can also explain seemingly nonsensical acts such as heroism and the opening examples from Florida. If we understand this motivation, we can use the knowledge to assist the suffering people who desperately need our aid but do not naturally inspire our urge to act—perhaps even Earth itself.

Even though I argue that we must broadly apply the science of behavior to understand altruism, I also consider information about how the act looks and feels to be relevant. The father of ethology, Niko Tinbergen, opined that in a desperate attempt to seem "scientific," physiologists had lost the forest for the trees. Studies proliferated about this cell or that nervous system tract, but they lacked information about the animal or the behavior in which the element was embedded. Tinbergen was hopeful that this problem would be resolved through an emerging field that would merge ethology and physiology, supporting rich behavioral descriptions with the underlying neurophysiology, but I am not sure his dream was realized.

Researchers who are studying altruism still try to seem more "scientific" (another term with a connotation that does not properly capture everything under it) by adding extensive control conditions to practicable laboratory experiments that produce statistically significant results. They focus on human acts that seem hard to explain but that are simultaneously easy to control, measure, and compare—not acts that are similar to how we give to one another in the real world or our ancient past.

Most current research on human altruism involves upper-middle-class, educated, white students who give some money from the experimenter to another university student for no reason whatsoever. The focus on money has become so eclipsing that editors sometimes refuse to publish an experiment that does not involve money, because economists disbelieve that an act is truly costly or measurable without it. Science is still losing the forest for the trees in this process. Researchers fail to understand or consider the most basic ways in which we care for one another. They do not consider acts such as emotional support or tight hugs as types of altruism, since these behaviors do not seem strange. (We are lucky indeed if our lives are so suffused with acts of care that they fail to warrant notice.)

We must flip this bias on its head. The fact that most of our aid takes the form of caring for those we are closest to does not render it common or unworthy; instead, it makes it real and important. It is the basis of human flourishing. Being emotionally and physically close to other people, in a way that feels good and raises our spirits, is so difficult and important that people seek therapy, for years or decades, to figure out how to do it better. Caregiving behaviors were essential to the very survival of our ancestors for hundreds of millions of years. They are an essential part of the origin of altruism—the origin of our species.

OVERVIEW

This book was designed to explain the altruistic urge, which we "instinctually" feel in very specific situations that mimic our ancestral (and still important) need to care for helpless offspring. I have described this altruistic urge enough times now that I have heard the same concerns repeatedly, primarily from people who take the view that altruism must be weird and special and must require a great, large brain (or an uncommonly giving spirit). As such, I focus this book on explaining the altruistic response model while addressing such common complaints. As a preview:

- The introduction provides a richly detailed description of the way that rodents that have recently given birth, which scientists call maternal dams, retrieve newborn pups back to the safety of the nest. This behavior is clearly adaptive, and we understand much about the neurobiology of this caregiving act, which greatly resembles altruistic responding. The remainder of the book argues that offspring retrievals like this are foundational to our understanding of altruistic responses, which can occur in similar types of situations.
- Chapter 1 provides an overview of the altruistic response model and a preview of the rest of the book.
- Chapter 2 explains how a behavior that we regard so highly in humans could possibly be shared with other species—rats, even—because of the way the brain itself evolved over millennia and shares features with the brains of other caregiving mammals.
- Chapter 3 explains how the altruistic response model applies to a specific form of altruism without necessarily affecting the types of altruism that we usually study, like deliberated aid or monetary donations. The altruistic urge is involved in these other forms at times, without being necessary or sufficient for them.
- Chapter 4 explains how something could be described as an "urge" or even instinct without meaning that the behavior is fixed, emerges intact, is context-free, or is relevant only to "primitive" species.

- Chapter 5 briefly summarizes the neural mechanisms supporting the altruistic urge and their overlap with known neural mechanisms for other reward-motivated behaviors.
- Chapters 6 and 7 address how the situation and qualities of the victim and observer influence the likelihood of an altruistic response, in a way that reflects the origin as a caregiving instinct.
- Chapter 8 describes the most popular existing theories for how altruism evolved, is motivated, and occurs in the brain. I compare and contrast the altruistic response model to these evolutionary and neuropsychological theories to demonstrate the benefits of this model within the landscape of other views.
- The conclusion reiterates the basic model, describing gaps in our knowledge and studies that still need to be done, and explaining why I intentionally did not extend this model to explain all of human morality.

Altruism is a lot of things to a lot of people. Herein I describe our origin as a species that is driven to help victims in the greatest need, without giving it a second thought. An impressive feat if I ever saw one.

ACKNOWLEDGMENTS

Thank you foremost to my family, who supported me with their love, wit, and independence for all of these months and years, as I took time out of our days and weekends together to work. Thank you to my mom and dad, who provided me with a loving and stimulating home for so many years . . . helping me to understand the meaning and value of caring for one another. I so appreciate my editors Eric Schwartz and Miranda Martin at Columbia University Press for supporting this work.

ABBREVIATIONS FOR NEUROANATOMIC REGIONS, NEUROPEPTIDES, AND NEUROTRANSMITTERS

Region	Acronym
Anterior cingulate cortex	ACC
Basolateral amygdala	BLA
Dopamine	DA
Dorsolateral prefrontal cortex	DLPFC
Hippocampus	HPP
Medial amygdala	MeA
Medial prefrontal cortex	mPFC
Medial preoptic area of the hypothalamus	MPOA
Nucleus accumbens	NAcc
Orbital frontal cortex	OFC
Oxytocin	OT
Prefrontal cortex	PFC
Paraventricular nucleus	PVN
Periaqueductal gray	PAG
Subgenual region of the anterior cingulate cortex	sgACC
Ventral pallidum	VP
Ventral tegmental area	VTA
Ventromedial prefrontal cortex	VMPFC

THE ALTRUISTIC URGE

INTRODUCTION

The Curious Case of the Assiduous Dams

I n a now-classic study conducted in 1969, a physiological psychologist at San Fernando Valley State College, William E. Wilsoncroft, was studying mother rats' motivation to retrieve their newborn pups. Before this study, research spanning back to the early twentieth century had already shown that such maternal rat "dams" are highly motivated to approach and contact their own pups. They will even retrieve the new pups of another dam to which they are not even related. This research documented that the instinct to retrieve pups was strong enough that the newly maternal dams were willing to learn complex mazes in order to gain access to young pups, and they even walked across electrified grids to reach their litters. The new moms were literally accepting electric shocks to access pups. Testifying to the relative strength of this motivation compared to other enticing rewards, dams were more willing to cross an even higher number of these electrified grids to access their litter of pups than to receive food, drink, or even sex. This instinct—referred to in scientific studies as "offspring retrieval"—is pronounced in these rats on the days immediately after giving birth.

In this particular 1969 experiment, Wilsoncroft was interested in determining if rat dams would press a bar to receive access to a pup that the experimenters delivered down a small chute into the test chamber, just as they would press a bar for food pellets—the more typical reward for bar-pressing in rat conditioning experiments. To measure the motivation of the dam to access and retrieve pups, Wilsoncroft initially trained the female rats when they were still pregnant to press a bar to receive food pellets so that they would understand how the system worked. Then, the day after giving birth, the dams were tested in a sequence that began with six bar presses, each rewarded with the original food pellets, followed by six bar presses rewarded with each dam's own pups, which the experimenters delivered down the same chute as the food pellets. Just as with the food pellets, the dams retrieved each pup from the conditioning chamber before carrying it back to their adjacent nest chamber by grabbing the pup by the "scruff" in their mouth, as is common in four-legged mammals (see figure 0.1). Afterward, the experimenters swapped each dam's own pups out for unrelated ones from another dam that had given birth around the same time. One after another, the dams pressed the bar to receive an unrelated pup down the chute, which they would dutifully return to the safety of the nest, before beginning the sequence anew (press bar, receive pup, carry to nest). The experimenters ostensibly created a circular conveyor belt of pups, which were repeatedly circled back to the eager dams, which were always ready for more. It is important to realize that, at this point in the experiment, the dams were no longer receiving any food rewards for acting, the pups were not related to her, and they were under no obligation to press the bar at all. The dams could have just sat there and rested once the food pellets or her offspring disappeared from the equation. The strange pups themselves were the dams' reward for pressing the bar.

The experiment lasted for three hours. The dams retrieved unrelated pup after pup after pup until the exhausted

FIGURE 0.1 Drawn depiction of how a rodent dam carries a pup in her mouth during a retrieval, usually back to the nest.

Stephanie D. Preston, "The Origins of Altruism in Offspring Care," *Psychological Bulletin* 139, no. 6 (2013): 1305–41, https://doi.org/10.1037/a0031755, published by APA and reprinted with permission, License Number 5085370791674 from 6/10/2021.

experimenters decided that the dams' pup retrieval response would not lessen. As they amusingly noted in the published article, "the only real extinction appeared to occur in [the experimenters] who got tired of removing the pups from the nest box and filling the delivery apparatus."[1] Wilsoncroft's figure shows how the typical dam retrieved approximately one pup every 30 seconds for the entire 180 minutes of the experiment (figure 0.2). Even more impressive, the best performing dam retrieved *twice as many* pups as the average dam—684 in total—traveling an estimated 4,000 feet in the process, half of the time (2,000 feet) while also dragging the pup in her mouth.

This study is evocative. It is just a short report, perhaps containing fewer total words than my description. It is memorable, of course, for the amusing imagery of pups traversing down a chute, into the nest, and back again—time after time, like kids

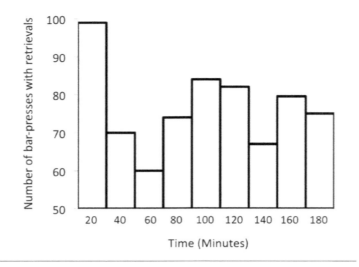

FIGURE 0.2 A histogram taken from the original 1969 Wilsoncroft study depicting the number of pup retrievals the dams performed over time until the experimenters gave up.

Stephanie D. Preston, "The Origins of Altruism in Offspring Care," *Psychological Bulletin* 139, no. 6 (2013): 1305–41, https://doi.org/10.1037/a0031755, published by APA and reprinted with permission, License Number 5085370791674 from 6/10/2021.

circling back to the top of the best slide at the water park. The article is also written in an older scientific style, whereby the researcher provided more conversational, vivid, and random details about the experiment than is permissible today. For example, reporting the specific quantitative feats performed by the most tenacious dam or revealing that the experimenters were the ones who became too tired to continue would be uncommon today, despite the fact that these attributes add considerably to our understanding of the phenomenon. Wilsoncroft was also mindful to report that no pups were hurt in the process—a sweet inclusion, given that animal cruelty was not particularly a hot button issue at the time, and perhaps a sign, in and of itself, of our concern for the welfare of neonates. Details like these, along with other odd descriptions from research at the time, such as the way confused dams tried to retrieve their own tails back to the nest, are essential to our ability to understand the biological and neural bases of the behavior. These details speak volumes about the hot motivational state and the almost fixed and reflexive nature of offspring retrieval, which is something that you could not just infer from summary statistics on the mean number of pups retrieved. Taken together, I love this old, brief report because it provides us with a strong and memorable sense of the sheer strength of the dams' motivation to access pups after birth—even those that do not belong to her, even without a reward.

I have chosen this study as a center point for this book, not only because it is amusing and instructive but also because it parallels a highly lauded but poorly understood human behavior: altruistic responding. This book describes how these parallels are not a mere accident or an analogy but rather reflect the fact that our very evolution as caregiving mammals motivates us to respond to others' urgent need with aid. The fact that we already possess so much data on the neurobiology of offspring care across species also allows us to build a more complete picture of human altruism, in the brain and in our behavior.

THE BENEFITS OF AN ODD EXPERIMENTAL DESIGN

When generalizing from the laboratory to real life, the Wilson-croft study is clearly an artificial situation that produced a more concentrated set of retrievals than would ever occur naturally in the wild. No rat dam in the field has ever had the opportunity to save or retrieve hundreds of unrelated pups in a single after-noon. We might observe her retrieving a handful of her own pups in the burrow or perhaps a few dozen if the dams lived commu-nally. As such, this experiment sacrifices some of the "ecological validity" of the behavior. However, the sheer extremity of their retrievals, which could only be observed under such artificial con-ditions, sheds light on the way that mammals evolved a power-ful motivation to secure pups. It is key for understanding the altruistic urge that we understand the true power of this moti-vation to retrieve—and its ability to reward the dams for doing so—which you could not infer from just observing a few natural pup retrievals in the wild. There are actually multiple attributes of this pup retrieval system that you would not assume from observing natural behavior, which are conveyed by this one infor-mative study.

If you observed a dam naturally retrieving her pup in the wild, you would not assume that she would also retrieve *unrelated* pups. It would be sensible to assume that such pup retrievals are restricted to offspring, given the privilege that evolution applies to behaviors that benefit one's own shared genes. However, the truth of the matter is more interesting and complex. Close sys-tematic observation and experimentation across species has dem-onstrated considerable variation in the degree that offspring care is restricted to related neonates or also generalizes to nonkin.[2] For example, if a species usually does not encounter unrelated neonates, there is no requirement for a neurobiological mecha-nism that discriminates between related and unrelated pups, such that one can avoid caring for nonrelatives. As a result, dams may be inspired to "accidentally" retrieve any neonate, just as they did in Wilsoncroft's lab. These accidents would usually not harm the species in the long run, since such opportunities are rare if not nonexistent. Conversely, if a species is surrounded by offspring that are related, as in many "alloparental" species that care for nieces, nephews, cousins, and other group members, then there is a genetic benefit to broadly offering care, owing to the shared genes and the mutual, reciprocating support provided by the group. Raising an infant by yourself in an isolated burrow seems very different from raising your infant in a group that provides mutual care but, in both cases, there is no biological pressure to evolve a mechanism that forces you to focus your efforts only on related offspring.

In yet another case, flocks of sheep give birth many unrelated lambs at the same time that must be nursed. Because nursing is so costly, and so many unrelated lambs are present at the same time, sheep evolved a sophisticated ability to immediately recognize and care for only their own lamb. Thus, only because of experiments like Wilsoncroft's and our ability to compare across species can we know that caregiving does not always depend upon relatedness. As a consequence, we also know that, as with human altruism, even vigorous care can be extended to strangers . . . under the right circumstances.

If you observed a dam naturally retrieving her pup in the wild, you might also assume that retrieval was encoded in the DNA of females. Beyond the question of gender, it would take considerable research to determine if females retrieve pups at *any* period in their lifetime, or only just after birthing their own pups. In fact, research described later in this book confirms that even virgin female rats and male rats will care for newborn pups with comforting, stimulating, and safekeeping behavior. These virgin female and male rats that are not in a "maternal" state do need time to habituate to the presence of such strange and novel pups to compensate for their lack of maternal hormones. However, once activated, even nonmaternal rats will care for pups, even unrelated ones, just as human females and males might assist complete strangers throughout their lifespan.

If you observed a dam naturally retrieving her pup in the wild, you would also not know whether retrieval evolved to respond to the *sound* of the distress call per se—in a necessary and sufficient fashion—or could be released in response to any reasonable cue of pup separation, danger, or distress (e.g., visual, olfactory, lost sensation). Dams surely perceive many aspects of the situation, but research shows that pups' ultrasonic distress cries are highly salient and motivating, just as the cries of distress and pain motivate us as humans to help infants and other people.

If you observed a dam naturally retrieving her pup in the wild, you might also assume that retrievals, as fixed and reflexive as they seem, are produced by a robotic motor program that is built into the DNA of rats. In fact, the retrieval of pups has been shown to be motivated by a hot motivational state that is similar to the drive to approach other desirable and rewarding consummatory objects, like delicious food or drugs of abuse. The motivation and rewards associated with offspring care engage the same brain areas that promote seemingly selfish drives, like the motivation of a cocaine addict to secure more drugs. Thus, rat dams are not simply enacting a genetically encoded, instinctual motor program. The dams are *compelled and motivated* to retrieve distressed and separated pups, just as we are sometimes compelled to secure that last slice of pizza or to help a stranger in need.

Taken together, even though the Wilsoncroft study is a little unrealistic, if we had only observed the natural sequence of pup retrieval in the wild, we would not properly interpret and might pointedly *misinterpret* how the behavior evolved and is supported by processes in the brain and body. That is part of why the

Wilsoncroft study is so elegant: because it reveals so much about the underlying mechanism in a single, well-designed, and maybe amusing afternoon in the laboratory.

WHY SHOULD WE CARE ABOUT RATS RETRIEVING PUPS?

The average person is not typically interested in rat neurobiology and would not see any link between the Wilsoncroft study and human altruism. But the case of our assiduous dams is critical for understanding the evolution and neurobiology of our own compulsion to retrieve perfect strangers in distress, danger, or need—and to feel good about it. This case of the assiduous dams is critical for understanding our *altruistic urge*.

People often create an artificial divide between humans and other species because we view ourselves as "special": as the endpoint of an imaginary, serial, evolutionary progression from simplicity to complexity. As such, we view the actions of rats as emerging from hardwired genetic programs that are enacted in a rule-based manner, whereas our helpful responses are thought to reflect a considered, rational choice. Such assumptions are particularly widespread when it comes to interpreting a lauded human behavior like altruism. At times, people will acknowledge that primates such as great apes might show primitive traces of care-based behaviors like altruism, but the buck stops there. After all, great apes look similar to us, and we hear that they are highly genetically related to us. Maybe dolphins also count since their brains are also highly encephalized. What they don't tell you in these reports and documentaries is that we are related to rodents by only 1 percent less than we are to great apes. We are genetically related to pumpkins by 75 percent. Thus, this genetic or aesthetic overlap is not the end-all of potential commonalities across species. Anthropocentric views of altruism grossly underestimate the degree to which even rodents possess biological mechanisms that are shared with humans and are complex, individually varying, and sensitive to context.

When people hear that rats have an instinct to retrieve pups or that people possess an *altruistic urge*, it can elicit the mistaken impression that the behavior is unintelligent and inflexible—and thus unlike our own rational choices to help. In reality, instincts are designed to be flexible and altered by features such as early development, individual differences, the victim's identity, and characteristics of the situation, as I describe later in this book. For example, even though the rodent dams were so assiduous in the experiment, that does not mean they would retrieve a pup in the presence of a predator, just as we would not approach a lost toddler at the mall if it would make us look like a kidnapper. The mechanism itself is designed to operate with minimal conscious deliberation while still being highly sensitive to context, because of the way that brains operate more generally. The mechanism is fast . . . and smart.

Of course, humans possess cognitive capacities that nonhuman animals do not.[3] We can perform, at times, great feats of intellectual and abstract reasoning, producing unique innovations like tall buildings and bridges, computer chips smaller than fingernails, and multinational charities that help starving people far away from us. Inversely, we share much of our biological heritage with other mammals and are thereby susceptible to some of the same instincts, particularly in situations that were important to our ancestors for millennia—like ensuring the safety, survival, and security of helpless neonates.

Thus, by deeply understanding how offspring retrieval evolved and is processed in the brain and body across contexts, even during strange experiments with rats, we can see how adaptive, complex, and sensitive these behaviors really are. This fact should alleviate some of the resistance to the idea that our own behavior is shared with that of other species, including rodents. The offspring care mechanism is, in fact, designed in most social species to permit the care of non-offspring while still ensuring that individuals do not endanger their survival or suffer from devoting excess time to care for nonrelatives. Thus, some "biological destiny" in the mechanism of offspring care is not a death knell for our free will or our insistence upon a flexible and sensitive system. The system itself includes an amazing amount of complexity—even in our rodent brethren—such that the "instinct" to care is sensible, conserved across species, and similarly issued toward other humans when we feel an urge to help.

SUMMARY

We can learn so much about the human instinct to help, from even Wilsoncroft's simple, engaging study of just five female rats that were trained to retrieve pups from a revolving chute. By attending to the important details of this study and integrating it with the extensive newer recent research on offspring care and human altruism, we can appreciate how this mammalian caregiving mechanism prepared us to respond altruistically, even to unrelated strangers.

The altruistic response model proposes that our own altruistic response to others in need derives from our ancestral need to protect helpless offspring, which we largely share with other caregiving mammals. This theory is informed by hundreds of research studies, but I focus the book on clarifying common concerns with the proposal. For example, I assure you that an altruistic urge does not mean that people are always helpful. Far from it. I clarify that the altruistic response model does not cover all human aid, but only a specific kind that is most similar to offspring retrieval. I describe how the altruistic urge differs across individuals and situations and is thus not an encapsulated and invariant response that is always issued. I explain how referring to a behavior as "instinctual" does not mean that it is reflexive in a maladaptive or thoughtless way, but rather in a way that inherently takes the individual and situation into account—even in rats.

The next chapter describes the altruistic response model in broad strokes, how it relates to other similar theories, and why it is important to examine at this time. This is followed by chapters that detail specific entailments of the model to assuage common concerns so that one can fully appreciate the theory. Once we understand the altruistic response model, we can better understand a very human and rational urge to rush toward those in need.

1

THE ALTRUISTIC
RESPONSE MODEL

The behavior of the assiduous dams from the introduction represents a precursor to our own altruistic response. The need to retrieve offspring evolved early in caregiving mammals, and this retrieval and caregiving response can be activated by adult strangers, which we refer to as "altruism." Caring for offspring is clearly adaptive because it promotes shared genes between the giver and receiver. No argument there. However, the instinct to retrieve helpless infants is built into our genes, brain, and body in a way that does not actually specify *who* we should help, only which stimuli we find motivating, under certain circumstances (e.g., in a parental state, when facing neonatal need). Because of this genetic inheritance, when we find ourselves in a situation that resembles infant care—when a helpless victim requires immediate aid that we can provide—an *altruistic urge* can be issued toward that stranger or even another species.

A situation that resembles a helpless infant usually includes a victim who is vulnerable, distressed, helpless, and in need of immediate aid that the observer can provide. These specific circumstances or requirements protect us against helping strangers who might be trying to manipulate us or can help themselves. A truly vulnerable and helpless other is most often a baby or child anyway, or an otherwise incapacitated individual. These stipulations also prevent us from acting urgently or with costly aid in situations that could resolve themselves or that do not demand our immediate attention.

In order for the altruistic urge to translate into an actual response (after all, we do not act upon *all* urges), the observer must also know the appropriate response and feel confident that they will succeed. The neural circuit for offspring retrieval in rodents prevents rats from retrieving pups when they are scared, intimidated, or uncertain. Similarly, humans fail to rush in when they do not know what to do or predict that they cannot help—or may even make things worse for the victim or themselves. This calculation of possible success can be achieved through implicit motor planning processes in the brain and does not need extensive conscious deliberation, even if our decisions are sometimes accompanied by conscious thoughts. Thus, there is a natural opponency between feeling an urge to rush toward victims who we can help and avoiding those we cannot, which explains our paradoxical capacity to be startlingly heroic as well as embarrassingly apathetic, despite having inherited an urge to help.

All of these features define the altruistic response model: a vulnerable, helpless victim in urgent need of aid that the observer can provide. Only when combined do these attributes transform a seemingly implausible generalization about human goodness into a scientifically supported argument about exactly when, why, and how people feel compelled to help, and when we sit idly by or even cause harm.

SIMILARITIES ACROSS CAREGIVING MAMMALS

Our urge to rush toward vulnerable targets in immediate need is considered comparable to the retrieval of pups in rodents. As such, scientists believe the altruistic urge evolved directly from the mammalian propensity to respond to neonates, which is similarly organized in the brain and behavior across species that share a common ancestor.

There are several ways in which I consider rodent offspring retrieval and human heroism similar. Both actual acts involve similar motor behaviors, occur under similar circumstances, and rely upon similar mechanisms in the brain. Importantly, both offspring retrieval and altruistic responding engage the same areas of the brain: specifically, regions in the hypothalamus coordinate with the mesolimbocortical system (e.g., amygdala, nucleus accumbens, subgenual cingulate cortex, prefrontal cortex) to drive individuals toward others who are distressed, helpless, or juvenile. Helping these distressed others, in turn, provides mother rats and us with a physiological sense of reward for our actions, which drives us to want to help again in the future.

In both pup retrieval and altruistic responding, the same neurotransmitters and neurohormones also modulate the response. For example, oxytocin lowers one's anxiety about approaching a victim and facilitates the bond between victim and care provider, while dopamine motivates individuals toward victims and makes the resulting close contact feel rewarding—which additionally feeds back to encourage future responses, as observed in our assiduous dams.

There are arguments in evolutionary neuroscience about whether some of the neural regions or labels that we use in this book are appropriate, such as the validity of concepts like a "limbic system" or "reptilian brain." I will discuss these issues in chapter 2, but this debate does not impact my central argument. Neural regions change over time and species in their exact location in the brain, structural form, and interconnections, but the similarities are sufficient to permit researchers to recognize the areas as "the same" in two species.[1] For example, a brain structure called the dopaminergic striatum has likely tracked rewards and motivated organisms toward valued items for millennia. The exact structure of this system—including where the receptors for dopamine are located, how many receptors there are, and which form they take—do differ over time and species to suit each species' needs and physical surroundings, but the region is still effectively the same and serves the same general function across species. The same goes for oxytocin, which has participated in the birthing process and offspring care for hundreds of millions of years. This hormone is effectively the same across mammals, and even similar in certain fish and birds. We expect some differences in the way a brain area operates across rats, humans, and other species, but the general principles hold, which are sufficient for the purpose of the altruistic response model.

DISENTANGLING TYPES OF ALTRUISM

In rodents, the offspring care system comprises both active and passive care. Passive care is defined as the nurturing that we stereotypically associate with females or being "caring," such as comforting, soothing, warming, and providing food and pleasant touch. Passive care in animal models is akin to the consolation described by primatologists Frans de Waal and Filippo Aureli in great apes, which have been observed consoling friends after a fight, something the primatologists construe as a form of empathy.[2] Passive care has already been linked to human altruism, with theories in psychology, biology, anthropology, and philosophy describing how the shared feelings and empathy between caregivers and offspring in early hominids or primates supports our general capacity for empathy, sympathy, and compassion—all of which can foster human altruism.[3]

The altruistic response model agrees that passive caregiving processes such as feeling empathy and soothing another can be activated by the victim's need and motivate aid. However, after taking this more incontrovertible fact for granted, the altruistic response model additionally explains why people sometimes provide more costly, active, and heroic aid, which is delivered more quickly than empathy-based altruism and does not require those intervening feeling states like sympathy. In rodents, pup retrieval and nest building are the "active" forms of care.[4] So much ink has already been spilled on the relation between passive offspring care in animal models and subjective affective states like empathy and sympathy that promote human helping, but little to no research has related active offspring care in animals to human altruistic responding (but see Michael Numan on cooperation).[5]

I fully acknowledge that active care does not represent or explain all forms of altruism. Most cases of altruism in animal biology result from a simpler formulation of shared genes that need not extend from offspring care. People sometimes also spend days or weeks making highly deliberated decisions about donating their money to just the right philanthropy. People also help acquaintances just because they want to get to know them better, to enjoy their company, or because they need their reciprocated help later. Sometimes people help just because someone asked them to or because they were taught to do the "right thing." All of these cases are forms of altruism, which are explained by other researchers in some detail; however, they are not "altruistic responding" as described herein. According to the altruistic response model, only forms of helping that stem from the motivation to rush toward helpless neonates are explained by the altruistic response model.

That being said, even if examples of altruistic responding in this book focus on concrete and immediate physical acts—the ones that are most like pup retrievals—the mechanism is often also involved in more abstract forms of altruism. For example, if you learn about a stranger's plight from a television commercial and are compelled to help by his or her vulnerability, distress, and immediate need of aid that you can provide, then even a financial donation that you took hours to decide upon would involve the altruistic urge, along with the other deliberated cognitive processes that are well known. People are biased to assume that only humans make altruistic decisions and that they do so consciously because such explicit and deliberated choices are the only type that we can directly observe and report on. But if the victim compelled you and pulled at your heartstrings (as it were), the response circuit that generated your initial drive to respond still strongly influenced the outcome, even if other factors were weighed in your mind. Because there is such a bias in the literature to assume that we help only through conscious processes, this book focuses on this ancient urge to help, without denying the existence of strategic, reasoned, or selfish forms of altruism that are well studied. The altruistic response is not defined by the type of action undertaken by the altruist but by features of the situation that motivated their aid.

THE NATURE OF AN INSTINCT

One of the reasons that people dislike a proposal like the altruistic response model is that they naturally recoil from the suggestion that humans possess an "instinct." The word brings to mind a genetically encoded behavior that cannot be controlled or moderated by the individual, context, or situation. This simple and lawful process is then restricted to nonhuman animals, given that we believe our choices are the product of a superior, rational mind that other species do not possess—especially rodents. Sharing an instinct with other mammals does not mean that we are the *same* as rats, it just means that we share a highly adaptive and necessary drive to protect our beloved, which sometimes leads to altruistic aid.

The altruistic response model does propose that humans possess an adaptive "instinct" to care for helpless, vulnerable others in need, when they can help. The instinct is even proposed to be linked through our shared ancestry with such seemingly disparate species as rats, mice, and monkeys. I even use the term "instinct" throughout this book, not shying away from this mischaracterization. But to address this misunderstanding, I devoted a full chapter to explaining how even mammalian "instincts" are embedded in epigenetic mechanisms that render them complex, context-dependent, and deeply influenced by experience rather than simple hard-and-fast rules. Even rats possess these rather sophisticated epigenetic mechanisms (in fact, we know of their existence from research on caregiving in rats). Thus, it is not the case that rats have a simple version of the retrieval instinct that is stupid and insensitive while we possess a complex and sensitive version. Both rats and humans possess biologically sophisticated "instincts" that are sensitive by design to the context of the individual's life and that can be overridden when it is irrational to respond.

The opposition between avoiding and approaching pups already prevents adults (rats or humans) from responding to helpless neonates if the situation seems too novel, aversive, or dangerous. For example, people are often unmoved by the need of individuals who are distant to them, compete with them for resources, or suffer from a problem that seems too big to resolve, even for victims that clearly need help. Thus, one of the strongest objections to a theory of human altruism—particularly an "urge" to respond—is the undeniable fact that we are *not* always helpful and kind. This is actually a benefit of the altruistic response model, because the design of the neural circuit that supports off-spring retrieval includes both an arm that avoids helping in uncertain or unsafe conditions (while inhibiting the approach response) and an approach arm that produces a motor-motivational urge to respond in situations that resemble offspring care (e.g., the victim is distressed, vulnerable, neotenous, helpless, and needs immediate aid that the observer can provide). This altruistic urge does not reflect a rose-colored version of reality that ignores our great intransigence; the natural opposition built into the neural system explains our paradoxical but adaptive capacity for hero-ism and indifference.

An urge also addresses individual differences, since people naturally vary in how they perceive the situation, such as the degree that the situation appears to resemble offspring need or that a successful response is predicted. Some people are inter-minably too scared to act and overestimate risk (e.g., in the face of anxiety or phobia), whereas others routinely rush toward sit-uations even when the deck seems stacked against them (e.g., with mania, when trying to impress, or when supremely skilled or capable). These individual differences, like the urge itself, reflect each person's genes and environment, which produce great vari-ation in the response.

Attributes of the situation that render it like an offspring retrieval (e.g., helplessness, vulnerability, distress, immediate need) are also not "all-or-none" requirements that are always present or pronounced. Each feature is present in the world in a continuous, independent, interdependent, and additive way that produces the strongest response when acting in concert; how-ever the attributes can substitute for one another and still pro-duce a response. For example, a victim who is unconscious on the subway tracks will not scream in distress, but we still realize that the victim is vulnerable because we understand the nature of unconsciousness as a form of intense vulnerability and need—so we feel the urge to help. Alternatively, a family member who screams in the house could cause us to rush across the house to save them, even if it turned out that they only stubbed their toe or twisted an ankle, situations that are not as urgent as an oncoming train but that still elicit a response.

The instinct is also influenced by your own past experience. For example, your response to a loud scream would be strongly diminished if you were at the amusement park but would be enhanced if someone were climbing the twenty-foot-tall rock wall at the park. People startle to loud sounds when they are walking through dark alleys or searching the house at night for an intruder, but not when they are reorganizing the kitchen cabinets or supervising a bustling toddler. This efficient and dynamic neural design, which evolved to be shaped by experience, allows us to appraise situations and their context quickly, during real emergencies, in ways that generally produce an adaptive response.

There are systematic biases built into the system that do sometimes cause problems. For example, because we are predisposed to see harm in cases like crying babies, we may underrespond to people with a traumatic head injury who just act strangely or remark on a headache in the absence of the telltale signs of injury like bleeding or crying. Conversely, we can overreact to individuals who resemble babies or juveniles. For example, episode 679 of the popular radio show *This American Life*, "Save the Girl," describes separate incidents where people's determination to save a sweet, young, innocent girl caused chaos and harm. Act 1 documents the story of a Vietnamese adult woman who was detained for over a year when entering America from Laos to meet her future husband because her childlike appearance caused people to insist that she was a child who needed protection from sex trafficking.

Our systems also include conflicting response tendencies that prevent us from acting when we might, such as our evolved tendency to avoid cues of contagious disease like grotesque injury, illness, or blood, even if the victim's specific problem is not transmissible (e.g., a bad arm or leg fracture cannot be transmitted to a helper, but our disgust can still inhibit an approach). Thus, our evolved instincts can lead us astray. People are also predictably biased to underrespond when their developmental, cultural, and personal experiences accumulate in a way that paints an altruistic response in a more costly light. The results of such biases are not always ideal, but they are not necessarily maladaptive, because our mammalian capacity to shift and change behavior with our environment is generally adaptive and often even benefits us.

As an example of the way our personal experiences can change us, when I first visited New York City without my parents, I was immediately approached by an overwrought man with a long story about how he needed money for gas after a long series of unfortunate events. I believed his story because of his convincing, overwrought appearance, and I gave him a whopping twenty dollars in sympathy. As soon as I walked away, I realized that his story was probably false, and I was less naïve going forward. Surely a real New Yorker wouldn't have been approached in the first place, but if so, he or she would probably not have stood there patiently while the stranger detailed his travails, and certainly would not have offered up such a large sum at the end. Similarly, when I visited Rome, my Roman friend was adamant that I pay attention to my backpack on the subway, keeping it to the front, close to my body. My mother had not received this lecture and was pickpocketed on the subway platform in Rome, losing all of her money, credit cards, and passport in one fell swoop.

After that, I was more cautious when traveling about, only carrying a little money at a time and keeping it somewhere less accessible. Conversely, my friend from India becomes uncomfortable when I engage with strangers in public because of her own early lessons with avoiding strangers who are begging for money. Despite these predictable biases, which we may learn in the span of a lifetime, it is still adaptive to possess an altruistic urge because our protective urge is important to survival and already sensitive to danger and context.

THE NEURAL BASES OF RETRIEVAL AND ALTRUISM

Research across species has examined the biological bases of offspring care. But most of our detailed neural and hormonal data come from experiments with laboratory rodents like our assiduous dams. Research techniques have advanced significantly since the early days, defined initially by behavioral conditioning or large brain lesions, since scientists can now measure gene expression, record from single cells, and stimulate brain areas in awake, mobile animals. Excellent reviews of this brain system already exist that were written by the primary researchers of these animal models of care,[6] and so for the purposes of this book I describe the system in broad strokes and emphasize the attributes that pertain to altruistic responding.

Active offspring retrieval is supported by a two-pronged neural circuit that supports both the avoidance and approach of offspring in an ancient brain circuit called the mesolimbocortical system, which usually includes the amygdala, nucleus accumbens (NAcc), medial prefrontal cortex (MPFC), hypothalamus (particularly the medial preoptic area, MPOA), and downstream

motor and autonomic regions that support a physical and physiological response (figure 1.1).[7] In brief, because of the opposing
neural circuits for avoiding and approaching, animals can switch
from a "default" mode of avoiding neonates to approaching and
caring for them when they are primed by the hormones of pregnancy and parturition or habituated to neonates. The avoidance
of aversive, novel pups is construed as the "default" response since
rodents that are not yet parents generally avoid pups. This avoidance arm of the circuit includes neural connections in the lower
portion of figure 1.1, proceeding from activation of the amygdala
to the anterior hypothalamus (AHN) and periaqueductal gray
(PAG) in the brainstem, which changes heart rate and supports
a fearful withdrawal from the neonate.

In contrast, when rodent dams are primed with the neurohormones of pregnancy and parturition, they become highly

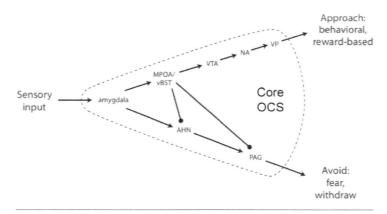

FIGURE 1.1 A depiction of the neural circuits that support offspring care,
from research on rodent pup retrieval, referred to collectively as the
offspring care system.

Stephanie D. Preston, "The Origins of Altruism in Offspring Care," *Psychological
Bulletin* 139, no. 6 (2013): 1305–41, https://doi.org/10.1037/a0031755, published by APA
and reprinted with permission, License Number 5085370791674 from 6/10/2021.

motivated to approach and care for pups, which inhibits that avoidance route (the amygdala inhibits the anterior hypothalamus) and activates the approach circuit represented in the upper portion of figure 1.1. In the approach circuit, the amygdala activates the MPOA of the hypothalamus and the ventral bed nucleus of the stria terminalis (vBST), which in turn activates the dopaminergic ventral striatal system. Even if all of these regions participate in a response, only the MPOA is considered essential to retrieve a pup, which inhibits the avoidance system and activates the reward-based approach system. Damage to other regions in this approach circuit can alter, change, or diminish the response, but only damage to the MPOA prevents it entirely. This necessary link to the MPOA is unfortunate for our ability to study this circuit in humans because the MPOA is very small and deep in the brain, which makes it hard to record from in people. Regardless, there is sufficient convergent evidence that the larger hypothalamus responds in humans to the cries of babies, particularly one's own.

Virgin females and male rodents are not already primed with the hormones of pregnancy and parturition but will come to care for pups after a gradual period of habituating to the pups, which is still supported by changes in the same brain areas and neurohormones. This is not so different from the situation in humans, wherein people without caregiving experience—even new fathers—often consider infants to be novel and frightening and shrink from responsibility if the mother is not nearby to assist or take over. Moreover, people generally do not find the truly helpless newborn babies to be as attractive as older ones, who seem less fragile. The infants that we find attractive on baby food jars are not actually newborns, but three- to six-month-old babies with the iconic features of a baby face like plump, rounded cheeks, a larger and rounder head and eyes, and pudgy, small limbs and noses. People even rate the newborns of other species as less cute than the slightly older versions.[8]

It is striking how similar true mammalian neonates resemble one another. For example, puppies and kittens who are newly emerged look like one another and also quite resemble rodent pups because they all have smooth, small, shriveled and largely hairless bodies and closed eyes. Human infants also possess a highly attractive smell that draws mothers to them. I tell my skeptical undergraduates that "newborn babies are delicious and smell like sweet, fresh peaches . . . you want to just 'eat them up!'" More than a metaphor, such phrases reveal the way that smell activates this offspring care system in much the same way that smells motivate us to approach more directly consumable rewards like delicious foods—because they all rely on that same neural system that motivates people to approach rewarding items of any type.[9]

The offspring retrieval sequence provides multiple evolutionary and genetic benefits to pups and dams. Keeping pups together in an underground nest reduces predation while providing them with nourishment, warmth, and stimulation, which helps them to grow and develop strong stress and immune systems. Thus, active and passive care combine to increase pups' fitness, and therefore also dams' fitness. Dams are also physiologically rewarded by the bonding and close contact with pups and their smell,[10] in much the same way that people can feel reassured and calmed by breastfeeding, hugging, and cuddling their young child. Thus, neonates are not inert objects that dams carry to and fro like so many bags of sand; they are engaging, rewarding, social partners that dams bond with and seek. These rewards are also critical for motivating caregivers to approach on subsequent occasions, teaching them the benefits of approaching that allow the behavior to continue after the intense hormones of pregnancy and parturition subside.

Taken together, the offspring retrieval sequence that was so amusingly depicted by William E. Wilsoncroft and colleagues is not just a random act or a fun little laboratory display. His description conveyed a very stereotyped and caricatured version of a natural, powerful, and rewarding instinct that is critical for the survival and flourishing of caregiving mammals under natural conditions.

It is important to understand that this "offspring care system" is not an independent entity in the brain that evolved or is committed only to this response. The mesolimbocortical system participates whenever organisms avoid aversive outcomes and approach desirable and rewarding ones, including when rodents and people observe food and drugs of abuse as well as attractive people, desired products, chocolate, wine, snacks, money . . . and, of course, babies.[11] This brain system is also engaged when rodents and humans acquire and amass food or goods, both of which they value and hoard under risky conditions.[12] This brain system may not have even originally evolved to support offspring retrieval, because long-term care of offspring may not have emerged until the late Triassic period, whereas the requirement to secure food and mates was ever present. Thus, the retrieval response may have piggybacked onto a system that already motivated behavior toward rewards in a fast and intuitive way.

The proposed homology maps these features of the rodent offspring retrieval system to our own responses to infants and the strangers who resemble them. We feel the urge to respond when we care about or are bonded to the victim and when we feel safe and confident enough to respond—which is supported by the approach arm of the offspring care circuit in the brain. But we are often hesitant to approach strangers in need, famously documented by the "bystander apathy" effect. We also feel good after rescuing a baby or stranger, since rescues usually end with reassuring close contact between the altruist and victim. For example, when a mother retrieves her crying child who has fallen off the swing set, there is usually a period of close comforting contact that is calming for both of them. This means that both the active and passive forms of care are inherently intertwined, as the former leads to the latter, which cycles back again to promote the former. Human research on altruism often focuses on the rewards of giving in the form of monetary reciprocation, the warm glow of giving, or increased status. However, at an ancient neurobiological level, we are also powerfully motivated by the implicit rewards provided by approaching and supporting someone in need, which was more consequential in our early evolutionary history.

CHARACTERISTICS OF VICTIMS THAT FACILITATE A RESPONSE

Because altruistic responding is derived from offspring retrieval, we are more likely to help adult strangers under conditions that mimic those of a helpless offspring, in urgent need of help that we can provide. Of course, babies are by their very nature vulnerable. They cannot do much of anything for themselves. They also possess "neotenous" features like large heads and eyes and shortened noses and limbs that are known to be compelling and attractive.[13] Babies also clearly display distress and need without self-consciousness or guile. This combination of vulnerability, helplessness, physical neoteny, and salient distress motivates our aid. Good caregivers do not respond to distressed infants when they get around to it, perhaps after finishing up an exciting book chapter or getting a few extra hours of shuteye. Successful infant care requires frequent, urgent responses to quickly changing needs, which is exhausting and wrecks the lives of new parents—but also explains when and why we help complete strangers. We help when the situation mimics that of a helpless offspring, when the victim genuinely seems helpless and legitimately needs immediate aid that the observer can provide and predicts will help.

CHARACTERISTICS OF THE OBSERVER THAT FACILITATE A RESPONSE

There are also features of the giver that predispose an altruistic response. There has already been extensive research on the degree that human altruism is fostered by empathic or altruistic personality "traits,"[14] to the point that I do not need to focus on them. It is assumed that prosocial personality traits can modify the altruistic urge. Adults do vary a lot in the degree that they feel for and want to help strangers. These individual differences derive from a host of factors that conspire in any given situation to predict the likelihood of an altruistic response. For example, empathy and altruism are altered by childhood experiences with empathic care and with one's worldview, concept of morality, and ability to regulate distress. Personality traits, however, are viewed as relatively stable within a person, and therefore they cannot explain why someone reacts in one situation but not another. For example, parents may be very empathic, calm, and patient with their children but short with other adults, which would not be captured by personality questions about how one feels about another's distress in general.

A key attribute of the altruistic response model is that responses depend on the degree that the observer predicts a successful response, based upon things like their expertise and competency (e.g., "Can I swim?" "Am I strong enough to yank the person back up?" "Is there time?" "Do I have enough money?") and the perception that the situation can be fixed (e.g., "Will my aid fix the situation?" "Is this a chronic problem?" "Is the victim resilient?"). Compared to other explanations of altruism, particularly those associated with empathy or personality traits, the altruistic response model uniquely emphasizes the observer's *motor* expertise. People rush to help, even under dangerous conditions, when their motor system implicitly predicts that the necessary response can be performed in time, successfully, without unduly injuring or endangering oneself. This calculation is performed quickly and effortlessly by the motor system, with little to no explicit reasoning. If the observer predicts that the response will not work, the pesky but very useful "avoidance" branch of the offspring care system will pervade, and help is not offered.

It can seem unfortunate or regrettable that we do not always respond to suffering. But these characteristics of the system reflect an important nervous system design that adaptively prevents individuals who are too weak, slow, or confused from rushing into ocean currents or burning buildings. The implicit prediction of success is not just reserved for highly physical or heroic cases. This prediction extends to mundane forms of altruism, like stopping on a busy city street to ask someone stooped against a building if they need help. A stranger on the street in need is like a new pup in that we are unsure of the potential risks and may overestimate the likelihood that they are aggressive, contagious, faking injury, or suffering from a problem we cannot address, such as an acute medical emergency, or a chronic condition such as schizophrenia or addiction. In reality, these issues are not usually as problematic as people predict, but, on the balance, our bias toward perceiving risk is generally adaptive. The implicit calculus that observers use to predict the success of their response prevents us from getting into trouble, but also fosters our embarrassing failures to act.

WHY WE MUST CONSIDER ALTRUISTIC RESPONDING NOW

The general proposition that our drive to help emerged from the need to care for vulnerable offspring is consistent with many existing theories of altruism, which is a good thing. Biologists, psychologists, philosophers, and anthropologists have assumed for decades, if not centuries, that the instinct to care for offspring was extended in evolutionary history from our own offspring to close, bonded, and interdependent group members, which benefited the helper and their group. For example, human empathy, sympathy, sharing, and helping have already been linked to the parental instinct by many well-known scholars from Hume to Darwin.[15] Recent books have also promoted this perspective, for example by Frans de Waal, Sarah Hrdy, Abigail Marsh, and others.[16] Thus, a general view of prosociality rooted in offspring care is not new and is largely accepted by most comparative psychologists. Even closer to this proposal, Stephanie Brown describes how this offspring care system from Michael Numan, which involves approach and avoidance, explains humans' costly, extended aid—such as when romantic partners, parents, and children sacrifice significantly to care for young, sick, or aging loved ones.[17] Numan similarly wrote about how his offspring care system can be extended to explain human cooperation.[18] Thus, others have extended the neurobiology of rodent offspring care to human altruism, making that element of the altruistic response model less controvertible. What then defines this proposal?

The altruistic response model is unique in the way it yokes altruistic responses per se to the form, function, and neurohormonal bases of offspring retrieval. Such specificity may seem misguided, inasmuch as scientists typically avoid being wrong by being just vague enough that they cannot be called out later when new data or logic emerge. However, the potential benefits of exploring this homology now outweigh the potential costs of relinquishing my plausible deniability.

I focus here on the parallel between offspring care and altruistic responding for multiple reasons. The way that dams retrieve pups is so functionally and structurally similar to active, heroic forms of altruism that we must at least explore the possibility that they emerged from the same mechanism. The opposition between avoiding and approaching pups also maps well on to the most popular explanations of altruism, which heretofore have not been unified: bystander apathy and empathy-based altruism. Both of these existing theories contributed to our understanding of human altruism, but they rarely intersect in the literature and sometimes even make opposing claims. Most theorists focus either on why we are apathetic toward strangers in groups versus helpful when we feel empathy more than distress toward victims that we value. The proposed homology between pup retrieval and altruistic responding explains both why sometimes we fail to act in a group even if we empathize with the victim and, conversely, respond even when we are in a group in a distressing situation. According to the altruistic response model, helping is not best explained by the number of bystanders or the level of distress (both of which contribute), but by a naturally predictive, dynamic, integrative neural process that spontaneously integrates the impact of others and one's own expertise into an implicit decision to respond.

The altruistic response model also places helping into a more ecological context than is typical for psychology, neuroscience, or economics, fields that often study altruism only through highly deliberated decisions to donate "house money." This proposed homology is all the more ecological because it emphasizes overt motor responses, which are often ignored despite being intrinsic to virtually all motivational processes, particularly those that were important to our early ancestors.

The altruistic response model is also important because it is the only theory to address heroism per se—the least understood form of altruism. Heroism is functionally and structurally distinct from most forms of altruism including alarm calls, grooming, food sharing, gift giving, and human donations of time or money. Heroism also occurs in stressful and arousing conditions that should theoretically undermine our urge to act. Heroism is also not addressed by evolutionary theories of altruism because it is usually directed at strangers who are unrelated and cannot return the favor. Heroism is also not consistent with empathy-based altruism, since heroes almost always report just rushing in without thinking, without the time to bask in feelings of empathy or sympathy or to imagine how the other must feel. Theories of sexual selection that propose that males evolved to be heroic in order to attract partners cannot explain why females outnumber males in virtually all forms of altruism except for the most physical forms and why we are biased to help young and vulnerable victims.[19] Heroes are surely praised and rewarded for their acts of bravery, but these rewards are also unlikely to be the primary or initial reason that we are motivated to respond. Mammals cared for dependent offspring long before they lived in large social groups or benefitted from advertising their beneficence to the crowd. Heroism is also not a "safe bet," since by definition it occurs under dangerous conditions that often kill or injure the would-be hero. A quarter of the Carnegie Medals for heroism are awarded posthumously, because the hero died in the effort

(90 percent are males). Thus, the altruistic response model can include social or sexual rewards from heroism as part of a larger explanatory picture, without assuming that these drivers were the original or most powerful factor in the evolution of an urge to respond.

Taken together, heroism does not fit well into existing explanations for empathy or altruism, including kin selection, reciprocal altruism, empathy-based altruism, sexual selection, or even general caregiving. Heroism is also difficult to study in the lab because it is intrinsically rare and physical. In contrast, most psychology and economic studies of altruism require the participant only to press a button to donate money to a hypothetical victim or unknown student for no reason whatsoever. Thus, heroism per se is almost never studied, apart from a few descriptive or phenomenological case studies of real heroes or how we view them— for example, on gender differences in heroism or altruists in World War II.[20] Because heroism has been difficult to explain, we have the most to gain by generating theories that cover active aid— both how it evolved and how it is processed in the brain and body—in a way that is easily integrated with more passive forms of aid and with existing theories of altruism such as inclusive fitness, reciprocity, bystander apathy, and empathy-based altruism. Because altruistic responding is so similar in form, function, and neurobiology to rodent offspring retrieval, we have much to gain from exploring this potential similarity.

To date, we really do not have any other neurobiological explanation for why people feel compelled to take extreme risks to help complete strangers. The altruistic response model explains this as well as more common and mundane acts of aid that are nonetheless important to daily life, such as holding the door for a stranger, helping an elderly person carry groceries, or donating money to feed starving children on another continent. The altruistic response model also predicts our inherent biases and apathy, which we can be used to increase aid where it is most needed—in the real world.

The altruistic response model aims to correct the imbalances in our understanding of altruism by considering how offspring retrieval explains human heroism in a way that is consistent with existing theories. Some of these ideas may turn out to be wrong. But by providing specific, testable hypotheses we can at least advance the field beyond where it too often lies: in vague speculation. Considerable evidence exists now for how caregiving operates across species and is comparable to human altruism. Drawing from this surfeit of evidence, we can augment theories about our empathic, sympathetic, and tender nature with theories about our proclivity for active, dangerous, and even heroic acts that help complete strangers.

SUMMARY

A homology between offspring care and human altruistic responding is proposed that relies upon the fact that there are shared neural and behavioral processes involved in how we approach, retrieve, and care for vulnerable others in need, across species and contexts.

The altruistic response model seems simple enough. It also accords with many integrative theories of human giving that already place parental care at the core of our capacity for empathy and altruism. Sometimes a view of altruism that is linked to caregiving just "feels right" to people, especially parents. However, the theory can also feel inherently problematic to people. People are nervous about extending a behavior associated with female parents to males and nonparents, they do not think of humans as possessing instincts, and they may not consider humans to be particularly nice. Because of the significance of these concerns, most of this book is devoted to addressing these potential pitfalls in order to reveal how an altruistic urge is sensible— despite and *because of* these concerns.

The chapters that follow largely parallel the structure of this summary, including chapters demonstrating the sensibility of proposing that offspring care and human altruistic responding reflect a shared common ancestor, the distinction between active and passive care and others forms of altruism, the nature of an instinct, the neural and hormonal systems that support the drive to help, the characteristics of victims and observers that determine when we approach, and how this model relates to existing views and is needed now. If the job is well done, those who are already comfortable with a care-based view of altruism across species will learn more specific and nuanced aspects of my framework, while the unconverted may come to see it as a reasonable, empirically supported view. Let us begin.

2

SIMILARITIES BETWEEN OFFSPRING CARE AND ALTRUISM ACROSS SPECIES

The fundamental assumption of the altruistic response model is that our behavior reflects an extremely long period of evolutionary history—hundreds of millions of years—in which our mammalian ancestors were preoccupied with ensuring their basic survival and that of their offspring. As such, in contrast to models of human behavior that focus on our "unique" abilities, I focus on what we *share* with other species.[1] Conscious cognitive processes like "theory of mind" and taking another's perspective surely play a role in human altruism. However, people emphasize these processes because they are so available to conscious awareness, and not because they have been demonstrated to explain more human behavior. Our urge to help in conditions that mimic offspring care is harder to observe because it was built into a primitive nervous system that profoundly influences behavior, even if we cannot witness its unfolding.

This chapter describes similarities between the retrieval of young pups and human altruistic responses that captivated me when I first read descriptions of offspring care in rodents. Our assiduous dams' retrievals of dozens of pups from the introduction and our own human rescues of strangers are similar in form, function, and neurohormonal bases.[2] These similarities suggest that we should at least consider that they are "homologous," meaning that they reflect a shared ancestry and are not just similar by coincidence but evolved separately.

ANALOGY VS. HOMOLOGY

This chapter explains the many ways that I consider offspring retrieval and care to be "homologous" with the human urge to help even perfect strangers. A "homology" in biology occurs when two body parts or processes in distant species exist because the two species share a common ancestor that also possessed this feature and passed it on to both subsequent species before splitting from one another. I argue that this is the case with retrieving offspring and altruistic responding. Because this homology is at the crux of my argument, I explain it in some detail in this chapter.

It could be the case that pup retrieval and human heroism *look* similar, but this resemblance is coincidental, convenient, or just poetic. Biologists use the term "analogy" or "analogous" when the two body parts or processes look similar in distinct species or genetic lines but did not emerge from a shared, common ancestor—perhaps arising independently in the distinct species because it was just a sensible solution to a similar problem that they both faced. For example, when you compare birds and bats, both have wings and can fly. The fact that birds and bats have four limbs is homologous, because both species derive from a shared ancestor that gave rise to the tetrapod line (as opposed to sharks and fish, for example). However, the actual structure of bat forelimbs is different from that of birds. Bat wings comprise skin stretched over spread finger and arm bones. In contrast, bird wings are made from feathers that spread along the axis of the arm, without involving any finger bones. Thus, even if birds and bats both have two wings and fly, they are assumed to have evolved these similarities independently, within their own branch of the genetic tree. This analogy leads to weird consequences like the fact that birds are actually genetically more like crocodiles and bats are more like rodents. Applied to altruistic responding, a homology is postulated between offspring retrieval and human heroism, with the assumption that retrieval and heroism not only *appear* similar but do so because of their shared mammalian ancestry. I will review the evidence for this assumption, with a particular emphasis on the fact that the same neural regions are implicated in both offspring care and human altruism.

SURFACE CHARACTERISTICS IN COMMON BETWEEN RETRIEVAL AND RESCUE

On the surface, the way that our assiduous dams quickly and urgently approached and retrieved helpless pups to return them to the nest bears a striking physical and functional resemblance to acts of human heroism. In both, a vulnerable, distressed, endangered individual in immediate need who displays distress is observed by someone who becomes compelled to approach them and return them from danger and back to safety. Both pup retrieval and human heroism also usually end with the rescued individual being held close by the rescuer after a place of safety is reached. The final close contact continues to protect the victim from harm while physiologically calming them both, which prevents damaging arousal and also rewards the giver who is then further motivated to approach in the future. For example, figure 2.1 depicts the real-life rescue of a three-year-old boy who fell into an

FIGURE 2.1 Drawn depiction taken from a video capturing
a female gorilla who retrieved a three-year-old boy who had
fallen into her zoo enclosure and carried him to safety,
protecting the boy from another gorilla.

Stephanie D. Preston, "The Origins of Altruism in Offspring Care," *Psychological Bulletin* 139, no. 6 (2013): 1305–41, https://doi.org/10.1037/a0031755, published by APA and reprinted with permission, License Number 5085370791674 from 6/10/2021.

enclosure at the Brookfield Zoo by a female gorilla named Binti
Jua. She was caring for her own baby when she picked up the
unconscious boy and rocked him as she carried him to safety,
protecting him from an elder female. Even though this is a gorilla
and a boy, rather than a rat and a pup or a human mother with a
baby, there is at least a clear structural and morphological resemblance to our own rescues and care of one another.

Pup retrieval and human heroism are also similarly described
as more of an urge than a rational, deliberated, considered

decision. The dams in William E. Wilsoncroft's study exhibit an intense and perhaps even irrational and somewhat fixed drive to retrieve the helpless, endangered pups, which did not fade away, at a time when new pups are maximally helpless.[3] The strength of this urge suggests a mechanism that evolved to address a strong pressure to secure helpless offspring, which fits with the way human heroes describe an "urge" to respond to victims in emergencies—even unrelated strangers.

Taken together, pup retrieval shares many superficial and functional features with human heroism: vulnerable distressed victims potentiate the response with a sense of urgency and immediacy that precipitates a physical rush to the victim, to retrieve them from danger to safety, ending with close contact that is calming and rewarding to both parties.

There are also profound similarities in mammalian social behavior that point to a homology in our behavior with that of other species. For example, as a research assistant in Tom Insel's laboratory, studying the neurohormonal basis of monogamous bonding in voles, I arranged social exchanges between animals and later coded videotapes of those interactions.[4] Subsequently, with Filippo Aureli and Frans de Waal at the Yerkes National Primate Research Center in Atlanta, I was charged with videotaping and coding the social interactions of rhesus macaque monkeys.[5] In both cases, it was easy to spontaneously identify the social script of a dominant and subordinate individual, which played out just as in a classic teen movie. There were always "bullies" who were more dominant, larger, more confident, and more likely to win in a fight. When the bullies approached less dominant, smaller, and less confident subordinate individuals, they could simply take whatever they wanted from the subordinate, be it their food or their mate. Dominant monkeys even displace subordinate monkeys from a random spot to sit on the ground that does not appear to have any special attributes; the act of displacing is itself seems to be the goal, which reinforces the social hierarchy. Subordinates spend much of their time avoiding dominants and quickly, voluntarily give up anything in contention to avoid a beating. Perhaps it is not just anthropomorphism to view these rodent or monkey scenes as similar to the bully in the teen movie who approaches a scrawny, nervous kid in the lunchroom and stares him down until the smaller one gives up his seat along with his precious food. These social dynamics not only look similar, but they are also subserved by similar neural and neurohormonal processes (e.g., involving the amygdala, hypothalamus, testosterone, vasopressin).[6] Thus, we know that much of social behavior is shared among rodents, nonhuman primates, and humans, which attests to a neural and behavioral homology in our social behavior more generally.

HOMOLOGOUS MAMMALIAN BRAINS

At the core, this altruistic response model is rooted in the belief that evolution preserves neural structures and functions over time. Mammalian brains were built over hundreds of millions of years. During this process, existing structures and functions constrained what could be built subsequently and how it could be implemented, just as the way that an existing home was built can constrain future renovations. Like a clever contractor, evolution elegantly and efficiently modifies what is already there and reuses existing structures to solve new problems. As such, neural processes that we inherited from our mammalian ancestors can powerfully affect our behavior, even when we cannot consciously observe the process.

It is hard to believe that we share much of our neural and cognitive processes with other species, especially rodents. After all,
our brains look very different. Rodent brains are tiny and
smooth—about the size of your fingertip. To a casual observer,
they are quite dissimilar from our larger, more rounded, and more
convoluted brains (figure 2.2). Rats also have a much smaller neocortex than humans, but they still have one, and it coordinates
with older areas like the amygdala and nucleus accumbens in a
similar way to promote decisions about avoiding or approaching
things that were previously pleasant or unpleasant.[7] If you are
still skeptical, consider the fact that all of our pharmacological
medications, such as ACE inhibitors and Prozac, are developed
and tested in rats and mice before being deemed safe and effective for humans. This would be pointless if our central nervous
systems were markedly different.

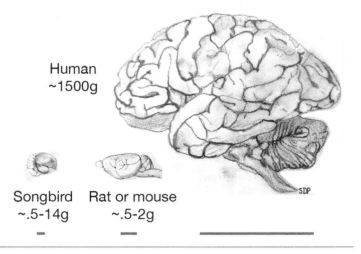

Human
~1500g

Songbird   Rat or mouse
~.5-14g   ~.5-2g

FIGURE 2.2 Drawn depiction comparing the relative size and complexity
of bird, rodent, and human brains, which highlights great size differences, despite highly similar brain areas and functions.

Sheep brains are superficially quite similar to our own, even if we think of sheep as very stupid indeed ("baaa!"). They are similar enough that biology teachers use them to teach kids about the human brain through observation and dissection exercises in the lab. Of course, the sheep brains that we dissect in neuroanatomy class are smaller than human brains—more like the size of a fist than a Nerf ball. But the structural and functional similarities are profound enough that much of what we know about the neurohormonal process of parturition (including pregnancy, giving birth, identifying one's offspring, nursing, and caregiving) derives from research with sheep.[8] We could not learn anything useful about human neural processes from rats or sheep if our nervous systems were so dissimilar.[9] Thus, even though there are superficial differences in the size or convolution of brains across mammals they contain virtually all of the same general areas, which perform similar functions and contain parallel interconnections, using the same neurotransmitters and hormones in order to process similar types of information—even between rats and humans.[10]

Research even shows that within the context of empathy and altruism, rats are not so different from us. Like people, rats and mice become stressed and aroused when they witness the pain or distress of a peer. As is the case with humans, this empathic, affective mirroring between the victim and an observer leads the observer to comfort and help the initial victim. Furthering this homology, rats and mice are more likely to help a peer under the same conditions that we do: when the victim is related, similar, familiar, or the observer has experienced a similar trauma and can relate to the type of distress. In one study, a region that is consistently activated in humans during shared, empathic pain (the anterior cingulate cortex) and the caregiving neurohormone oxytocin were involved when a monogamous prairie vole comforted a familiar, stressed cagemate (but not a stranger), just as those processes support the retrieval of a pup in rats and human altruism.[11] Rodents and humans not only have fairly similar brains, but they also respond similarly to others' suffering and need—evidenced both when they respond and when they do not—through the same neural regions and neurohormones that support caregiving and human prosociality.[12]

It is not inconceivable that aspects of this homology extend even further back in the evolution of neurohormonal systems. For example, even ants will rescue their peers from entrapment, in complex and clever ways that are altered by how similar the other ant is, the type of trap, and the relevant solution for untangling them.[13] Skepticism is warranted, and I am hesitant to assert that these ant rescues are truly homologous with human heroism. But they definitely look and function similarly and share an underlying mechanism (i.e., hormones released by the trapped ant that the passing ant perceives, causing a contagious stress response in the observer, who then responds).[14] The contagious stress response in ants mirrors the contagious stress that we have demonstrated in humans, wherein a human who becomes stressed and releases the hormone cortisol during a stressful event (a public speech that includes difficult arithmetic) also elicits a cortisol release in empathic observers.[15] Thus, either employing contagious stress to promote aid predates the evolution of the central nervous system or there are severe limitations on how to solve this problem, which produces similar solutions independently in each taxon (an analogy rather than homology).

The fact that I believe that humans have homologous brain systems and functions to other mammals does not preclude humans from having unique qualities. I am surely grateful that humans are capable of great feats of engineering and architecture as I sit in a warm house on a cold winter day in

Michigan, observing the birds and squirrels that seem miserably huddled under a branch against the cold, driving rain. But focusing too much on our human "specialness" leads people to overlook considerable similarities. Important skills that ensure survival are "baked in" to the mammalian brain. For example, brains are exceedingly good at learning from experience to avoid punishment, seek rewards, and integrate perceptions and sensations to make adaptive choices.[16] This inheritance means that much of our behavior can be mediated by undeliberated but still complex computations that are shared with other species. As such, our decisions are more influenced by prior affective experiences than we want to believe, and, conversely, other species are capable of more sophisticated processes than we understand.

Nonhuman animals can even surpass humans when their ecology demands a specialized skill. For example, the cerebellum, which is associated with motor behavior and motor learning, is larger in cetaceans such as dolphins and orcas to help them navigate long distances in three dimensions (figure 2.3).[17] People are often willing to acknowledge that dolphins are weirdly intelligent for a seafaring nonprimate, which they explain by the fact that dolphins live in social groups as primates do. But even small rodents that are not on our list of "smart" species (e.g., chimpanzees, bonobos, dolphins, and ravens) can possess neural specializations that yield amazing behaviors. For example, the Merriam's kangaroo rat has a larger hippocampus than highly related species like the banner-tailed kangaroo rat. The smaller Merriam's kangaroo rat cannot protect its stored seeds against larger, more powerful species that can just raid their den and make off with months' worth of foraging effort in the desert basin. To deal with this, Merriam's kangaroo rats evolved a specialized ability to cache seeds in small packets all over the

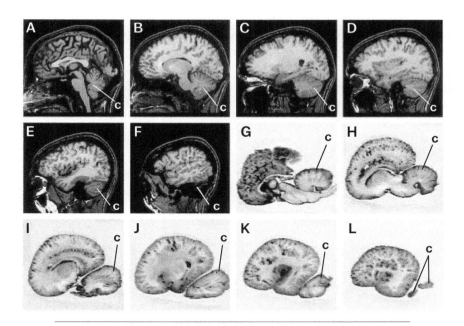

FIGURE 2.3 Images of the cerebellum in humans and bottlenose dolphins, demonstrating a relative size increase the region supporting spatial navigation and memory in the aquatic species, which swims in three dimensions.

Figure 1 in Lori Marino et al., "Relative Volume of the Cerebellum in Dolphins and Comparison with Anthropoid Primates," *Brain, Behavior and Evolution* 56, no. 4 (2000): 204–11, https://doi.org/10.1159/000047205, with permission from Karger Publishers, CC-BY-SA-4.0 License 5073741342007, 5/21/2020. The final, published version of this article is available at https://www.karger.com/?doi=10.1159/000047205.

desert—the location of which they remember so that they can efficiently dig the "scattercaches" back up during a drought.[18] Similarly, food-storing birds like Clark's nutcrackers live in cold climates at high elevations where it would be hard to find seed caches that are covered in snow all winter, as they are in the Rocky Mountains. During the peak caching season in fall, these birds are estimated to be able to cache up to five hundred seeds per hour, up to fifteen miles away, and to remember the location of up to ten thousand seeds in a single winter so that they can survive the long, unproductive season. Like food-storing kangaroo rats, Clark's nutcrackers also possess a larger hippocampus (also called the dorsomedial cortex), which has more neurons in it relative to the total forebrain volume and body size of corvid birds that do not store food, like the brash California scrub jay.[19]

The existence of a specialized hippocampus in a mammal and bird is profoundly important for the argument of a neural homology, because the clade that gave rise to mammals split from birds more than 300 million years ago. Thus, in both mammals and birds there is an identifiable brain structure, which resides in a relatively similar spatial location in the brain, which mediates spatial learning and specializes when needed to help each species acquire food, within their unique ecological context. The hippocampus is the same region that is affected in human Alzheimer's disease, which causes profound memory loss. The avian hippocampus is considered homologous with the mammalian and human form because it has similar attributes in cellular composition, arrangement, interconnections, neurochemistry, and cell types—even in taxa that split from one another long ago. Hippocampal specializations to store food are thought to have evolved independently from an ancestor that did not store food (e.g., the common ancestor before the Corvidae split from the Paridae did not store food), and species within each family independently evolved this specialization. This analogy does not conflict with the belief that the hippocampus itself is largely homologous, in the sense that when spatial memory specialization does need to emerge in vertebrates, it consistently relies upon this same brain area rather than just randomly assigning the task to any existing structure or a completely novel region invented for that purpose. Instead, evolution builds upon cells and functions in the medial pallidum that are already capable of performing this task—over and over again.

SHARED PHYSIOLOGY BETWEEN RETRIEVAL AND RESCUE

The human brain and nervous system evolved over hundreds of millions of years in the mammalian line to promote some specific behaviors, including avoiding danger, acquiring mates and food, and raising offspring. For the most part, all of this is accomplished without the benefit of explicit, conscious deliberation. Any time a new behavior or capacity emerges in our repertoire, it reuses existing genetic, neural, and hormonal mechanisms to achieve that goal, usually through minor changes in the way that the same genes are expressed.[20] Thus, you could "invent" a new way to address a new ecological problem just as you could invent a great castle to replace your small cottage, but in both cases, it is usually more efficient and effective to just make minor modifications that solve the problem reasonably well. Both brains and houses are situated in preexisting conditions that strongly constrain the solution.

As an example of such neural reuse, much of the system for retrieving offspring relies on the mesolimbocortical system, which acts through neurohormones and neurotransmitters such as dopamine, oxytocin, vasopressin, opiates, and others. The mesolimbocortical system is not specific to our drive to acquire bonded offspring. It is involved any time organisms are driven toward something that is predicted to be pleasurable or rewarding from experience. These rewards extend from newborn babies to alcohol to drugs of abuse and even to beautiful, expensive purses and shoes that humans invented only recently.[21] Privileged people in industrialized Western societies may take for granted that rewards like food, water, mates, and offspring are available, but these items are required for survival and are not always available in nature. Thus, the mesolimbocortical system employs fast-learning and highly motivating processes that ensure that mammals detect, remember, and seek such essentials. This system is also engaged when people make decisions to help other people, even when the decision requires a modern and abstract form of aid like money.[22]

Some aspects of offspring care are not the same as what is more generally required to approach other motivating rewards. For example, offspring care may be uniquely heightened by the process of becoming pregnant and giving birth and may uniquely require activity in a very specific hypothalamic nucleus (the medial preoptic area or MPOA).[23] Even so, most of the system is shared across species and contexts. Many scientists agree that the mesolimbocortical system participates in both caregiving and some human prosocial behavior, but only the altruistic response model links pup retrieval specifically to altruistic responding. This association may be too specific for many readers to stomach. But the response does appear to extend beyond the context of new moms caring for their neonates.

CAREGIVING EXTENDS BEYOND DAMS

It seems problematic to a theory of heroism that the best responders in animal models are postpartum dams, whereas human altruism is performed by a broad range of individuals— and most heroes are men. Moreover, students who are not yet parents and older adults are often generous even if they are not in a peripartum state. For the homology to hold up, the neurohormonal mechanism of offspring care must occur in other contexts.

This is not a problem for the altruistic response model, because even the foundational retrieval of pups by rodents is issued by male and virgin female rats as well as the dams that just gave birth.[24] Usually, these nonparental individuals first avoid the novel pups, which apparently emit a salient and perhaps aversive smell. Occasionally, a nonparental male even tries to eat the helpless pup.[25] I witnessed this firsthand when working with monogamous voles (figure 2.4). Voles are rodents, like rats, but they are good animal models of human monogamy because we can compare the brains and behavior of monogamous versus nonmonogamous vole species (just as we compared rodents and birds that did and did not store food earlier) to determine what must be added. My job was to compare the behavior of unmated virgin

FIGURE 2.4 Drawn depiction of biparental voles, which bond to one another and care for offspring together, through neurohormonal mechanisms that overlap with the offspring care system.

male voles toward an unrelated pup to their behavior after mating with a bonded partner.[26] The monogamous prairie voles clearly and visibly shifted behavior: before mating, they avoided pups in the opposite corner of their enclosure; after mating, they approached and even huddled with and groomed the strange pups, just as our assiduous dams did. I did sometimes have to terminate a pre-mating session when an aggressive, unmated male looked as if it were starting to eat the pup. But the shift to caregiving after mating was striking. (In a fun follow-up study, we tried to control for the visual characteristics of the pups as small, brown, oblong objects by replacing them with Tootsie Rolls. Not a single male tried to approach, huddle, or eat these candies, even though they were technically the only edible option, which perhaps says more about these candies as food than about their similarity to pups. But I digress.)

Males and virgin female rodents can be shifted into a parental mode in multiple ways, all of which reflect the avoidance versus approach opponency of the underlying offspring care system. Females can shift into caring if we artificially administer the hormones that females typically experience during pregnancy and parturition,[27] further indicating the hormones' causal role in offspring care. Even males and virgin females can be shifted into caregiving if they are simply allowed to habituate to the novel pups, as first shown by Jay Rosenblatt in the 1960s (and which has been replicated hundreds of times since), or if we apply the hormones associated with maternal care to these nonparental animals.[28] Rosenblatt simply left these nonparental individuals in contact with unrelated pups for an extended period of time. Over the course of about a week, the nonmaternal rats shifted from avoiding the pups to being less actively avoidant, to curiously approaching pups, finally providing them with the same nurturing care that a mother would. This transition is not so different from the way new human parents and babysitters can take time to become comfortable with newborns, eventually finding their footing, creating a bond, and then developing into confident caregivers. This caregiving transition in rats is all the more impressive since rats are not monogamous, and males and virgin females do not usually jointly care for infants in the wild.

CAREGIVING EXTENDS BEYOND RODENTS

So far, we have focused on research from the common laboratory rat. Even if motivational processes are similar in rodents and humans, a mammalian homology should be visible in species other than rodents. There are indeed studies in sheep, monkeys, humans, and even birds that reveal similar neural, hormonal, and behavioral processes for caregiving across species. Even nonmammalian species far from us on the evolutionary tree such as squid, crocodiles, clownfish, and rattlesnakes are known to sequester and protect their young from predators during their offspring's earliest and most vulnerable period, just as the rat dams did.[29] Males in monogamous species such as prairie voles and marmosets and tamarins also show increases in oxytocin, vasopressin, and prolactin and decreases in testosterone when they become fathers and care for offspring, just as our retrieving rat males did.[30] These changes do not require gestation and birth but involve a combination of cues from the mate, the infants, and prior parenting experience.

The precise specifications of the mechanism do differ somewhat across species to fit the ecological context.[31] For example, ewes quickly learn the exact identity of their own lamb to avoid nursing one of the many other lambs born around the same time.[32] Primates such as monkeys, apes, and humans may provide care earlier in development, as juveniles, because of the way that social groups benefit from alloparental care and that "babysitting" can prime processes needed later to care for their own offspring.[33] But differences across species or breeding conditions do not compromise the proposed homology, because the same relevant neural regions and neurohormones participate in the process, even if the exact number and location of receptors change in each instance. As with the spatial memory of food-storing animals, it has been proposed that alloparental or paternal care of offspring arises independently when needed.[34] However, I consider these cases to be homologous because they arise in the same neural regions and implicate the same neurohormones, which only require small changes to things such as the way the genes are transcribed, rather than a brand-new process.

CAREGIVING IS PROVIDED BY MALES

Multiple scientists who study caregiving have shown that human fathers experience similar hormonal and behavioral changes after the birth of a child, despite not experiencing the pregnancy, birth, or breastfeeding that support maternal care in females.[35] Human males show increases in progesterone and reductions in testosterone after becoming fathers. Males may require more habituation to be motivated toward the infant, because they are usually not exposed to intense hormonal changes as is the peripartum female, but when changes do take place, they occur through modifications to the same underlying system.

Males also become more responsive to infant distress when they become fathers, as mothers do, and this predisposition is retained thereafter: that is, once an experienced father, always an experienced father. For example, human fathers held a doll for longer and were more concerned by infant crying when they had greater prolactin and lower testosterone.[36] In a similar study, fathers' sympathy and desire to help an infant was higher in men who were already fathers than those who were not, and their response increased with prolactin and decreased with testosterone (although fathers increased testosterone during infant crying).[37] Reductions in testosterone may represent a shift toward nurturance and sociality, but these hormonal responses are sensitive to the context. For example, fathers still launch a significant testosterone response when they need to be protective or competitive but can be tender and nurturing when needed by momentarily suppressing testosterone and increasing estrogen.[38]

Of course, in humans, men and women can differ in the ways that they help, with men perhaps being more overtly physical and protective, supported more by vasopressin or testosterone than oxytocin. For example, male recipients of the heroism award from the Carnegie Hero Fund Commission more often saved a stranger than someone they knew, and the victims were more often vulnerable individuals who were very young or elderly.[39] Men often possess greater size, strength, and speed, and thus the altruistic response model presumes that men are more likely to perceive emergencies that require such skills to be manageable, activating the approach over the avoidance arm of the neural circuit. At times men are also more driven to act through testosterone, which is associated with physicality, risk-seeking, and the drive to display one's prowess. For example, the "life history theory" of Martin Daly and Margo Wilson argues that changes in testosterone in males evolved to promote displays of strength, power, and expertise in order to attract potential mates and outcompete rivals (which perversely also increases their risk of dying young).[40]

Taking the life history theory further, some researchers posit that heroism evolved to help men be noticed, praised, and selected as a higher-quality mate.[41] Our heroes do receive widespread public attention, medals of honor, public ceremonies, and even cash prizes. For example, Wesley Autrey became famous after rescuing a student who fell onto the subway tracks in New York after a seizure.[42] He received a public ceremony, was honored by the mayor of New York, received a $10,000 reward, and was featured in nearly every newspaper and major media outlet in the nation. Wesley Autrey later became the poster-man for academic treatments of heroism. Thus, being a hero in a highly physical and dangerous situation does seem to come with rewards, although media attention is arguably not always pleasant. The fact that these rewards for heroism exist, however, does not mean that a hero like Autrey acted *for* those rewards or that those rewards were the original evolutionary reason for his response. The need to display a valued behavior like heroism to one's group members in order to curry favor or mates was probably not relevant until relatively recent human history, when we started living in increasingly large social groups; in contrast, females and their male partners or family members would have needed to protect offspring to maturity long before this. In our current context, the social and mating benefits of seeming heroic can exist and contribute to the benefit of an altruistic response, but they are considered less powerful or ancient than the caregiving benefits. Even more relevant to this specific altruistic response model, Wesley Autrey may have been an ideal observer as a father who also reported expertise with tight spaces (i.e., he predicted that he would fit with the young man in the narrow space between the tracks and train as it went overhead) owing to his union job, which involved fitting into just such constricted places.

CAREGIVING IS PROVIDED TO NONKIN

Even Wilsoncroft's dams were willing to retrieve unrelated pups, and so the behavior can clearly be applied outside of one's own related offspring. Across species and examples, even if the simple math suggests that we should prefer to help our own offspring and other related individuals, there are many cases where individuals care for nonrelatives. This is not necessarily maladaptive because this extension of care occurs in situations where there is no strong selection pressure to avoid such care, and there is even some benefit to the self from being helpful.

Even though rats and mice *can* identify relatives—usually through smell and familiarity, a mechanism that also prevents mating with close relatives—there was likely no strong selection pressure to avoid caring for nonrelatives in rodents, which permits their similar response to unrelated pups. Rodents are "hider species" that live in underground burrows where contact with unrelated neonates outside of their social group is unlikely. Thus, rodents might only exhibit this behavior in unusual, artificial, and unlikely situations like Wilsoncroft's experiment. The fact that the mechanism does not prevent the care of nonkin does not mean that altruistic responses are maladaptive, given that these odd circumstances rarely occur in nature, making exploitation unlikely.

Some species that live in cooperative, social, or even "eusocial" groups (such as ants) provide care for others as part of their social structure. Such social structures are assumed to benefit the group as a whole because the group can share an unpredictable supply of food or defend against predators through safety in numbers. For example, when prairie dogs view a dangerous predator, they make an alarm call to warn the rest of the group, which signals the others to retreat to the safety of their burrows. If their social groups were smaller, fewer individuals would be available to detect a predator, and more of them would be killed. If the group were less interrelated, the benefit to saving them while risking yourself by being so noticeable would decrease. Monkeys also alarm call to alert group members of danger. Monkeys in the savanna even make distinct calls for different predators so that group members know, for example, to look up and duck into the grass if it is a hawk or look at the ground and run if it is a snake.[43] When individuals depend on the actions of others to survive, the mechanism is less strict about what it responds to, even if the species is technically capable of recognizing and bonding with their specific offspring.

Social primates may be more naturally caring than Wilson-croft's rodents, because they provide care even before they enter a parental state that involves significant hormonal and brain changes.[44] For example, caregiving in monkeys is provided by juveniles, unrelated virgin females, and sometimes males. Despite this, maternal hormones are still important and initiate and augment females' interest, care, protection, and treatment of infants. For example, pigtail macaque monkeys care for infants even when they are not parents, but pregnant females that are bathed in the parental hormones show even more interest in infants and provide more care even before giving birth, particularly if estrogen was added.[45] Apes also help in ways that are similar to an instinctive retrieval, such as by spontaneously helping an experimenter pick up a dropped object.[46] Old World langur female monkeys, which are considered less sophisticated or similar to humans than great apes, show an intense interest in newborns across stages of life, for example by responding to the calls of infants and attempting to hold and carry them.[47]

Humans are also social, group-living species who provide care well outside of the family unit. People raise unrelated foster and adopted children whom they have not carried or birthed, even when they do not know the biological parents. Nonetheless, adoptive parents form a strong parental bond with their children and care for them throughout life. Thus, humans also provide care for both related and unrelated individuals. Many social animals provide care even when they are not new mothers, and an altruistic response that derives from offspring retrieval and the attraction to neonates is still possible in other species, breeding conditions, and ages—in males and females alike.

Future research should confirm that people are more likely to approach nonrelatives under conditions predicted by the altruistic response model: when the victim is young, helpless, vulnerable, distressed, and in need of aid that the observer can give. For example, human observers in a bystander paradigm are hypothesized to avoid approaching a distressed, strange infant if anyone more qualified, related, or familiar is present. People often fear that they will do the wrong thing, make the situation worse, or become chastised for interfering when it was "none of their business." This fear, supported by the avoidance arm of the neural system, prevents people from acting even if they are concerned for the victim, which is why empathy does not always produce altruism. But children are often separated from their parents in public places and, at some point, are offered help by a stranger. According to the altruistic response model, strangers will approach when they are experienced caregivers, attuned to distress, and less concerned about how others perceive them. Children probably have complementary strategies for seeking nonthreatening strangers to ask for help, which suggests that this dynamic goes both ways, although this has yet to be tested. Of course, these are not easy studies—some even come with significant ethical problems—but there are ways to simulate such situations and approximate the context of a dam retrieving her helpless pup. We have already confirmed in our own research that people are more willing to donate money to victims who are infants or children over adults, particularly when the aid is needed immediately and involves some nurturance.[48] Thus, the existing evidence supports the model, but more is needed.

SUMMARY

To accept the idea that altruistic responding evolved from our inherited need to care for helpless offspring, scientists needed to demonstrate that there are homologous neural and behavioral systems that support such care across species. This burden of proof falls particularly hard on the altruistic response model because it is so tightly yoked to the retrieval of new pups by rodent dams. We are not rats. And we are not particularly impressed when someone helps their own child. For such reasons, this chapter was designed to demonstrate that pup retrieval and altruistic responding are similar, at multiple levels, despite these potential pitfalls. For example:

- At a structural and functional level, both a dam retrieving her pup and a human retrieving a stranger from danger involve a helpless, endangered individual who requires an immediate rescue from danger.
- The brains of humans are similar to those of other mammals, including rodents. Brain or region size or neurotransmitter receptors shift with species' ecology, but the general

existence, relative position, interconnection, and function are still similar, even when the specialization arises independently in two species after splitting from a common ancestor. Thus, it is reasonable to assume that we share neural systems and circuits with other mammals, including rodents—*homologies* that have been extensively documented in the context of caregiving.

- The same neural regions and neurohormones support offspring care across species and support altruism in humans (e.g., amygdala, nucleus accumbens, anterior cingulate cortex, prefrontal cortex, and oxytocin).

- Rodent pup retrieval is not just restricted to dams. Offspring care is also demonstrated in virgin female and male rodents and relies upon the same basic mechanisms (e.g., changes in maternal hormones, with bonding, and after habituation to pups).

- The mechanisms of offspring care are not restricted to rats or people; these processes similarly support care in mice, voles, sheep, monkeys, humans—even to some extent birds and fish.

- Human males and fathers also care for infants, through similar neurohormonal changes that shift them from being less engaged to concerned and responsive. Thus, despite a focus on the way that rat dams retrieve pups, this model is assumed to apply to humans and to males.

- Even if caregiving is usually reserved for related offspring and altruistic responding is specific to strangers, the mechanism has been shown to support care to other social partners, including nonrelatives, in multiple species. Social primates provide care for other group members, facilitating altruistic responses toward strangers under the right conditions.

If we take all of these points of connection into account, it is more likely that the observed similarities between dams retrieving pups and humans rushing into burning buildings or icy waters represent a homology. These behaviors are similar because they evolved from a common need in mammals to care for helpless, slowly developing offspring in urgent need of help to ensure their very survival. These behaviors are subserved by similar neural and neurohomonal mechanisms but can be modified to suit the ecological needs of the individual, gender, developmental period, and species.

3

DIFFERENT KINDS
OF ALTRUISM

The altruistic response model is focused on extending the active offspring retrieval that has been well-described in rodents to explain human altruistic responding. This focus was selected because, from a scientific perspective, we currently know the least about what motivates people to provide this immediate, active form of aid. But human helping is diverse and occurs in many forms. Thus, a theory of altruistic responding cannot and should not try to explain *all* types of altruism but should parse them on the basis of their evolutionary origins and shared processes, within the brain and body.

The altruistic response model is yoked to a specific behavior: the retrieval of helpless neonates by rodent dams. This analogy applies most overtly to heroic physical rescues that share both the form and function of a dam's rush toward helpless pups so that they can be retrieved back to safety. Understanding how offspring retrieval relates to human heroism would still be useful even if it turned out not to apply to other types of altruism, since researchers currently have almost no theories for how heroism evolved and occurs in the brain and body. The altruistic response model, however, does also explain broader forms of aid because of the focus on a powerful motivational state that can be activated in observers when the conditions mimic that of off-spring care. As such, aid can still be facilitated in human contexts that do not seem heroic or active but that still involve a distressed, vulnerable victim who needs immediate aid that the observer can provide. This motivation can even participate in people's decisions to make abstract financial donations to victims they cannot directly encounter, as long as these precipitating conditions exist. The altruistic response is defined not by the form of the act, but by the observer's underlying motivational state. This distinction allows us to separate forms of aid that evolved at different times, and that are supported by different neural and behavioral mechanisms.

OTHER CLASSIFICATION SCHEMES FOR ALTRUISM IN PSYCHOLOGY

I define altruistic responding as any form of helping that applies when the giver feels motivated to assist a vulnerable target after perceiving their distress and immediate need.

This definition already captures a wide range of altruistic responses beyond simple rescues or retrievals, but then it excludes forms of altruism that emerge from different evolutionary, motivational, or neurohormonal processes. For example, when people help to follow social norms, to achieve longer-term strategic goals (like cozying up to a powerful neighbor or boss), or to impress others, their aid would not count as an altruistic response, even if it were helpful and costly. Those types of aid involve higher-level goals or plans that motivate aid rather than a genuine and immediate urge to address a victim's clear vulnerability, distress, or need.

Altruistic responding excludes aid that emerged later in evolution and that relies upon a more complex mix of cognitive processes. For example, cooperating with your group to achieve a long-term goal like hunting, warfare, or building a warm structure requires many cognitive and neuropsychological processes that extend over a long period of time, which are not required for a simple rescue. Cooperation and its requisite cognitive processes did not necessarily evolve to promote caregiving per se, even though may people describe their theories of cooperation as if they represent all forms of altruism—at least in humans.

Cooperation and strategic aid *can* involve altruistic responding when the act is initiated by the perception of a victim who resembles a neonate (i.e., helpless, distressed, needing immediate help that we can provide). For example, you might stop to help a man that is stranded by the roadside next to his inoperative car because his plight simply motivated you to pull over, in which case, an element of the altruistic response participated in the act. But the same act of aid may not involve this urge, for example if you stopped because you felt guilty that you previously failed to fix his car at your auto shop, wanted to impress your passenger with your infinite kindness, or remembered lessons from your parents about doing the right thing. In those latter cases, the aid entails other motives and considerations that are not linked to caregiving, but these considerations can coexist with, augment, or limit an observer's response, particularly when there is time to decide. Thus, there are multiple routes to altruism and these routes can combine in varying degrees to increase or decrease the response, depending on the situation. Whereas it might feel neater and tidier to have a priori rules for classifying acts under a taxonomy of types of altruism (e.g., if there is a retrieval, then it is an altruistic response; if there is hubris, then it is sexual selection), in reality that is not how the brain works. The brain inherently integrates a variety of types of information, in a continuous, connectionist, and associative way, which allows for bits of this and bits of that to all work in tandem to produce an adaptive response—often outside of our awareness.

Placing "altruistic responding" into context with other classification schemes, this active aid is similar to Felix Warneken and Michael Tomasello's overt "helping."[1] Helping is defined as an overt behavior that has been documented even in young children and social mammals such as apes, dogs, and dolphins; therefore, helping is thought to be a more primitive behavior than forms of altruism that do not exist outside of human adults. The way that young children and nonhuman animals can help one another can involve the altruistic urge, but this should not be assumed. Conversely, you cannot assume that all altruistic responses look like overt help, particularly in humans. For example, you may be inspired by witnessing the distress and need of starving orphans across the globe and then write a check that takes weeks or months to assist orphans who are not even the same ones that you observed. This response would be undergirded by an altruistic urge even if it did not look like a direct, overt act of helping. Conversely, you may hold the door for someone struggling with packages because you perceived her clear distress and need or simply because you wanted to seem polite or were taught to be a "good person"—or any combination of these reasons, plus many more. Only if your act was motivated by the recipient's need would it count as an altruistic response, even if all cases of door holding involve a direct, overt act. For example, we performed experiments on campus where real people were observed holding or not holding the door for a researcher who posed as a student that seemed happy or sad. People actually held the door more for happy than sad people, even if the latter individuals seemed in greater need of aid; people did increase aid toward sad people when the researcher wore a bandage outside of a medical clinic, and they helped sad hospital patients more when they were asked to donate money rather than to sit with the distressed patients.[2] Thus, many motivations can inspire us to help, which are not clear from the nature of the act. This classification of overt help is useful in the context in which it was developed—to explain nonhuman aid—if you assume that these contrasting motives to avoid guilt, or to seem good, or to be "moral" are not relevant in those species. However, affiliative and strategic motives have been documented in other species and cannot be distinguished from caregiving motivations on the basis of the act alone.

Altruistic responding also overlaps somewhat with Frans de Waal's "directed altruism."[3] De Waal created this category to capture altruism that is directed at one specific individual, in order to separate direct aid like ape consolation from cases like alarm calls that have fewer cognitive requirements and that exist in many more species. For example, a prairie dog or monkey could risk its own survival by alerting the group to a predator, which increases the success of the group but is not directed at any one individual. The canonical case of altruistic responding is like directed altruism, with one individual directly assisting another. But there are cases where an altruistic response can help many individuals or when something could look like directed altruism that is not motivated by the other's distress or need. For example, in 2018 the world watched television coverage of a soccer team of Thai boys who were trapped in an underwater cave. Some of those viewers were inspired by the boys' plight and so wired money across the world to assist with the rescue attempt. These donations would count as altruistic responses because they were motivated by the boys' helplessness, vulnerability, distress, and immediate need; however, they are not good examples of either "helping" or "directed altruism" because the donors gave abstract aid that is not possible in other species, the distant helpers did not perform the rescue, and the recipients were a group of boys and not a single individual. Conversely, directed altruism could seem like an altruistic response but may not qualify when the motivation differs—like the motorist who pulled over to help a stranger to impress their passenger. Helping and directed aid overlap considerably in the types of altruism they describe, and both are useful for segregating more recent nonhuman primate aid from the basic inclusive fitness that exists in many more species; however, these classifications cannot help us segregate forms of human aid that are subserved by different underlying neurophysiological processes that emerged at different points in mammalian evolution, but that look similar on the outside.

In still another classification scheme, Kristen Dunfield divides prosocial behavior into helping, sharing, and comforting, because these three emerged at different periods of development and are subserved by different sociocognitive and neural processes.[4] This scheme is more like my own in design, because it aims to link the aid to the ultimate and proximate mechanism. However, this scheme is also a poor fit for the altruistic response because it focuses on one's mental capacity to discern the other's needs, which is more pronounced in humans, rather than on a more primitive motor-motivational state that is shared across species (e.g., to rush to comfort someone or pull them back from danger). The Dunfield scheme has some similar requirements to know, enact, and be motivated to issue the act; however, like the former two schemes, it was designed to explain the emergence of different forms of altruism in human development and thus does not cover altruistic responses that exist across species, could emerge early in development, and could cut across helping, sharing, and comforting (e.g., all three would be involved if a vampire bat shared food with a juvenile that failed to forage and needed sustenance and warm contact to survive a cold night).

Existing classifications of altruism all have validity in the context in which they were devised, for example, to deem acts as common or uncommon across species or stages of life. In apes or young children it is difficult to track motivation, underlying physiology, or cognitive process during helping, but it is easy to distinguish whether they helped an experimenter grab an item, shared a branch, or fed a child. My own requirement is to provide a classification that defines altruistic acts by when they evolved, for which purpose, and through which neural and physiological process. Altruistic responding can participate in abstract, distant, or distributed forms of aid that emerged later in evolution and development, as long as the act was inspired by the altruistic urge. There is no existing category of altruism that suits my need to distill this underlying process, which was also pointed out by scientists who were studying the rescue behavior of ants that I described earlier.[5]

Anne McGuire devised categories of altruism by studying the characteristics of helping that people report and associate with different types of events.[6] She found that most human altruism was either just casual, substantial, or emotional or that it involved emergency aid, which people judged by the perceived benefits, frequency, and costs. This scheme does involve the motivation for helping and is one of the only to include emergency aid; however, like the former schemes, it categorizes behavior by how the act appears to us from the outside rather than the relevant motivation or evolution of the act. For example, someone could provide any one of these types of aid (casual, substantial, emotional, or emergency) because they were motivated by the others' helpless or babylike state. The obsession over judging whether someone's aid was costly or "truly altruistic" is important to laypeople and to scientists who need to separate human altruism deriving from empathy from the less lauded need to signal one's goodness to others; however, this obsession is irrelevant to the goal of understanding biological predispositions that required an evolutionary benefit in order to emerge and that have persisted in the genome for so long.

In social psychology, scientists use a different scheme to classify aid, which again reflects the goals of that field. Notably, C. Daniel Batson spent his career demonstrating that people were capable of helping others through a truly altruistic and other-oriented concern for the other (i.e., sympathy) and not just a selfish desire to reduce their own caught sadness or distress from the victim.[7] This division was necessary to contradict the assumptions of many theorists that "true altruism" does not exist, since all helping can benefit the giver, who may only be acting selfishly. This division between Batson's "empathic concern" and "personal distress" is still a focus of research today. Even though this is one of the only schemes to focus on the helper's underlying motivation, it cannot capture altruistic responding because people can feel either empathy or distress from a victim's helpless need, which can propel them to act, even if they report feeling differently. Observers in most of these experiments also have time to soak in and think about their subjective state and to decide what to do about a situation that is typically not urgent. As such, sometimes people help without empathic concern and with distress and, conversely, feeling empathic concern does not always precipitate action. The altruistic response model is designed specifically to cover this gap, when empathy does not foster action and action occurs without empathy, based on factors such as the degree of immediacy, expertise, or time to consider one's own competing goals.

ACTIVE VS. PASSIVE CARE IN RODENT CAREGIVING

Scientists who describe offspring care in rodents refer to two forms: "passive" and "active." The term passive care was intended to describe the more nurturing and succorant types of aid that are generally performed within the nest, like huddling with, nursing, licking, and grooming pups. Active care is reserved for two specific behaviors that require more energy and leaving the safety of the nest: nest building and pup retrieval.[8] These classifications are a little confusing: nursing is considered passive but is energetically very expensive to the dam, and nest building and pup retrieval do not seem very similar to one another. Nest building is an anticipatory act with important long-term consequences that is performed only when one's immediate needs are met (e.g., to feed and be safe), whereas pup retrieval solves an immediate problem that redirects attention and energy from less urgent, longer-term concerns. Pups are also much more motivating and rewarding than nesting material, although both involve a retrieval-like act, so it is interesting to consider their overlap.

The mechanisms of passive and active care also overlap significantly, which we fail to appreciate if we always separate them. Both passive and active care occur in new mothers and are promoted by a cascade of neural, hormonal, and behavioral changes that support pregnancy, parturition, and caregiving. Moreover, the same neuropeptides (e.g., oxytocin and vasopressin) support both active care behaviors, and competent caregivers must do both active and passive care well. Passive and active care also both involve approaching an individual that is otherwise aversive and, thus, entail an inner conflict between avoiding and approaching the one in need, which is overcome under the right circumstances. Both passive and active care also produce positive rewards that increase the behavior in the future. Thus, at a conceptual, evolutionary, and mechanistic level, passive and active care overlap considerably, and you would not want to consider them totally distinct (see figure 3.1). Despite this, I focus on the active care form of pup retrieval because this act requires additional features that are not required

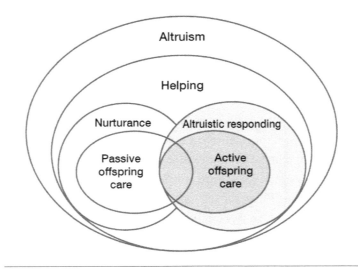

FIGURE 3.1 Venn diagram demonstrating different types of helping, some of which (but not all) relying upon the offspring care system, which itself includes passive and active forms—the latter of which are the focus of this book.

Stephanie D. Preston, "The Origins of Altruism in Offspring Care," *Psychological Bulletin* 139, no. 6 (2013): 1305–41, https://doi.org/10.1037/a0031755, published by APA and reprinted with permission, License Number 5085370791674 from 6/10/2021.

for passive care and that people have yet to apply to human altruism. For example, active care in rodents requires a specific region in the hypothalamus and a specific motor act and motivation that helps us understand why people who feel empathy may not always help or why sometimes people do help even when they are aroused, stressed, or distressed. Active care is also beneficial for explaining heroism, which was heretofore hard to integrate with models of empathy-based altruism that describe more passive forms of aid like consolation.

I use the term "altruistic responding" throughout this book instead of "active care" because it covers literal retrievals such as heroic rescues, as well as less physical forms of aid that are motivated by the same underlying circuitry. The term "altruistic response" also avoids the problem of artificially segregating passive and active care, which rely upon much of the same neutral and hormonal processes. For example, someone might donate money to "save" strangers observed on a television commercial, who seemed vulnerable and in immediate need, even if the act of donating isn't really an "active" physical response and the donor does not perform the actual rescue. Monetary donations like this still count as an altruistic response if they are motivated by a context and victim like that of offspring care.

As an example of this process, we performed a brain-imaging experiment at Michigan with Brian Vickers, Rachael Seidler, Brent Stansfield, and Daniel Weissman. In this experiment, participants were asked to read a variety of descriptions of fictional charities and then were offered the opportunity to donate any amount of money they had earned in preceding finger-tapping trials to support these victims.[9] The participants could not actually "rescue" anyone, and information about the victims was delivered through short written descriptions of each charity. Even though such prompts are not as salient as witnessing an actual rescue, participants still donated significantly more of their money to charities that helped babies or children who needed immediate, nurturant aid. For example, babies in intensive care or children being rescued from abusive homes were more compelling to people than adults who had suffered from an avalanche or a capsized boat that needed to be rescued. The additional motivation to help young victims was associated with brain activity in regions that help plan and generate motor responses, interpreted as evidence that this motivation is a hot, bodily response that primes you to approach such sympathetic targets. Returning to the concept of an altruistic response, I apply this term and process to such cases that are motivated by a context resembling offspring care, even when there is no actual immediate, physical rescue.

PASSIVE CARE IN THE SCIENCE OF ALTRUISM

There is not currently a dichotomy in the science of altruism between passive and active care. Researchers do often assume that our sensitivity to others' distress derives from the need to be sensitive and empathic toward our own infants.[10] For example, Frans de Waal and Filippo Aureli have examined how one primate consoles another group member that was injured or upset after a fight, or how they might reconcile after a fight through hugging or grooming.[11] Consolation has been documented in many species, including chimpanzees, gorillas, and bonobos, but perhaps not monkeys. Macaque monkey mothers do not even seem to reassure their own offspring after a fight, suggesting that their focused nurturance is restricted to the early neonatal stage. In more than two thousand anecdotal reports of empathy or altruism in nonhuman primates, researchers found that apes but not monkeys were observed comforting a distressed group member.[12] Note that even if mammalian consolation occurs only in the larger-brained great apes, the act itself does not appear to require a large neocortex, given that it also occurs in ravens from a completely different taxon, with much smaller brains.[13] Ravens actually express many social acts like primates, including forming bonds, breeding in pairs, and a slower development, which is inferred as evidence that group life explains more about these behaviors than raw brain size. Taken together, passive, nurturing forms of care are observed across species and have been linked theoretically to caregiving but they are not classified by anyone as "passive care" the way they are in the rodent caregiving literature; moreover, this nurturance in primates is not distinguished from the more active forms of aid such as rescues and is not linked per se to the offspring care circuit (although individual regions in this system are implicated, such as the amygdala and anterior cingulate cortex).

Social psychology and neuroscience do not study either passive or active care in segregated or prototypical forms, since research subjects in a controlled setting usually just donate money indirectly to people who are not present, and sometimes do not even need help. Common sense tells us that people do indeed exhibit passive care, like giving a reassuring hug to an upset friend, snuggling with a tired child, or wrapping a blanket around a boater pulled from icy waters. Such tender, comforting behaviors are important in daily life, especially in close social relationships where their presence almost defines the quality of the relationship.[14] Developmental psychologists measure passive care, such as when a child in the home comforts a distressed parent or experimenter.[15] As early as the first year, children exhibit some forms of help in these feigned situations, like hugging or patting the other. Children also intermix passive comforting aid with active help, such as bringing the sad parent an object that is needed or will make them feel better.

Perhaps by design, passive care and consolation are not only beneficial for the victim but also for the helper and the group as a whole. The consoler and the consoled can calm down more quickly after a distressing event if they are together than alone, minimizing the long-term consequences of stress on the nervous system.[16] This two-way link between comforting the other and yourself at the same time is central to maternal care and is even observed in human children, albeit sometimes awkwardly. When my youngest child used to fall down or get upset, her older siblings would practically headlock her in an intense effort to soothe her, trying to calm and please themselves while the youngest struggled against their "ministrations." Frans de Waal similarly described scenes of spreading distress in baby macaque monkeys that would run and jump onto one another in a big pile, seeking mutual comfort from the precipitating event.[17] Other researchers have delved even more deeply into this phenomenon to show how this physiological linkage, which emerged from the mother-offspring bond, reduces stress and links helping to improved health.[18] Sometimes people view this bidirectional reward as evidence that people are not really altruistic because they can be viewed as helping only to assuage their own discomfort. This is a short-sighted view given that the mechanism must provide some benefit to the altruist in order to exist as an evolutionarily stable strategy. This mechanism adaptively helps the most vulnerable, fosters bonding and positive states, reduces negative states, and encourages observers to approach again—it must also be powerful to override the uncomfortable and potentially dangerous feelings associated with approaching someone who is distressed.

Some have argued that acts of passive care such as hugging, patting, or grooming should not be considered altruism because they are not costly to the giver. I disagree. All physical aid carries some energetic and opportunity costs and renders you less vigilant to threats from the outside while you help. Helping is also very risky socially. In nonhuman primates, if one consoles the loser of a fight, the dominant, winning individual may attack and punish the consoler for allying with the loser. The close contact required by consolation also includes great social and emotional risks, inasmuch as such intimacy is welcome only in very limited relationships and conditions (as you well know if you have ever been rebuffed while trying to comfort or snuggle up to someone). The emotional punishment in such situations is very real indeed. People can also resent being consoled because it makes them feel patronized, infantilized, subjugated, or embarrassed—reasonable responses given that the behavior evolved to soothe infants. Finally, giving hugs is no less costly than other forms of altruism that we study in the lab, like passing a dollar or two of someone else's money to a stranger or picking up a few pencils or papers that the experimenter dropped. Even if you donated thousands of dollars to someone, it would not be particularly costly unless—like the poor woman in the Bible—you parted with money that you truly needed to survive . . . unless it hurt. Taken together, passive care does indeed carry with it very real costs and, despite being common and normative, should be taken seriously as primitive and important expressions of aid. It seems natural to observe a parent comforting a child, as depicted in figure 3.2, but just imagine enacting this same passive consolation with an upset stranger or crying friend—or even the crying child of your friend. The fact that people rarely display such distress and almost never provide such care outside of the offspring context attests to the costs of doing so.

FIGURE 3.2 Black and white rendering of an original painting Madonna with Child, by Esther Anna (Mattson) Stansfield, which depicts the common close, warm, bonding, and rewarding contact between mothers and their offspring.

With permission from R. Jon Stansfield.

We need more research on passive care in humans, but this is not easy. Most research subjects are strangers who will not want to touch or come close to each other. One could apply paradigms from relationship and caregiving science in which familiar dyads interact in the laboratory in seminatural situations, measuring the reassuring touch or responses to the partner's pain.[19] India Morrison and others have even measured specific nerve fibers in the skin that selectively respond to the slow, soothing, and caressing forms of touch, which link soothing behaviors to physiological rewards.[20] Kathleen Vohs and others have measured how close people place their chairs to one another in the laboratory after a stressor to indicate care or consolation. In our lab, people did not move their chairs closer to stressed experimental partners, but they did say supportive or reassuring things. Close contact may be so forbidden between strangers in our culture that even placing chairs near one another may be too weird. Creativity is required to examine passive care so that we can determine if this distinction with active care is useful for understanding human altruism. Studying passive care is also important as one of the most common and influential ways that people help each other in daily life, certainly much more common than heroic acts.

ACTIVE CARE IN THE SCIENCE OF ALTRUISM

I argued earlier that it is relevant and important to understand passive care in both rodents and humans. However, this book is focused on extending *active* care to our own altruistic responding. I chose this focus for a few reasons. To date, active and heroic forms of aid have mainly been studied from a phenomenological perspective, such as through case studies of heroes or through self-reports about what makes someone a hero.[21] Active aid has not been specifically addressed by evolutionary or neuroscientific theories of altruism, even those that assume a basis in offspring care, because these prior models focus on our sensitivity to, compassion for, and resonance with others' distress, without regard for the type of aid that follows or how the act is mediated in the brain and body. Studies on bystander apathy do measure a physical and active approach toward someone in need that is an active form of care,[22] but this is usually done to demonstrate why we *fail* to help rather than when we feel the urge to help; bystander paradigms also rarely include important features of an altruistic response, such as costly aid, rescues, danger, or expertise. Thus, as was true of passive care, existing research on active care does exist in some forms, in some fields, but it is not really studied as a form in its own right, which is different from passive care or may be analogous to pup retrieval or rescues per se.

SUMMARY

By carving nature at its joints, and aggregating only types of altruism that derive from similar evolutionary, motivational, neural, and physiological processes, we can understand a powerful mechanism that evolved over millennia to promote the sensitive, protective care of those closest to us—those who truly need our help. By merging information across disciplines and levels of analysis, in a single framework, to explain particular forms of altruism, we can merge previously disparate theories of altruism into a single, larger framework.

4

WHAT IS AN INSTINCT?

The idea that we evolved a natural capacity to help others may seem far-fetched, particularly because it describes an instinct or *urge* to help. Let's look at how we can possess such an instinct or urge that is nonetheless flexible and sensitive to context, just like altruism itself.

Altruistic responses are "instinctual," but that doesn't mean they're senseless acts that we cannot control or that they are the same across individuals and contexts. Rather, the majority of behaviors, instincts even, are encoded in a sophisticated nervous system that, by design, is sensitive to the individual's own genes, early life, environment, and current situation in a way that is flexible and largely adaptive—even in rats. Thus, we can have an instinct that sensibly varies by person and situation and urges are usually inhibited when they are disadvantageous.

The strict and nonexistent division between nature and nurture is like a Platonic form that is incredibly hard to defeat. Nearly every time a scientist is interviewed about a new finding, she is asked, "so . . . is this innate or is it something we learn?" Recently, the *Chicago Tribune* framed the results of a study about altruism in mice by Peggy Mason and colleagues as fodder for the nature/nurture debate.[1] (Who is really having this debate, anyway?) After almost every talk I give on empathy, someone asks if empathy is innate or if people can learn it. (Yes, and . . .) When I was in graduate school, an *APA Monitor* article title by Beth Azar captured this point in such a pithy way that we have used her title to make this point ever since: "Nature, Nurture: Not Mutually Exclusive."[2] The stickiness of this division may reflect the oppositional way that people in Western culture tend to think, by viewing things in black or white or as incompatible opposites, in contrast to an East Asian concept of yin and yang that comprehends the peaceful coexistence of opposites.[3]

At one time, scientists did actually believe that at least some behaviors were "hardwired" in the rigid sense of being encoded in the animal's DNA and released without training, perfectly encapsulated, in the absence of external input. However, evidence had already accumulated by the latter half of the twentieth century to correct this view. For example, when I was a student, we watched films about "feral children" like "the Wild Child," who was "raised by wolves," as depicted in a François Truffaut film of the same name.[4] We also watched a film about Genie, who spent her early years chained to a potty chair by her deranged parents, even more disturbing because her case occurred almost two hundred years later, in twentieth-century America.[5] Genie was housed from the elements and was fed, but she was completely deprived of normal psychosocial and linguistic interaction during her early development. Both Victor of Aveyron and Genie spoke and even moved in bizarre ways that seemed more animalistic to observers than human, which suggested to psychologists that we may be hardwired to walk, run, and talk, but even so require proper developmental conditions for these basic capacities to typically unfold. It was virtually impossible for the new caregivers of Victor or Genie to teach them adult-level language and integrate them into society, even if they were technically still children when discovered. Based on these seminatural demonstrations, and many more controlled ones, researchers began to realize that language requires a "critical period" for development, within which time relevant external stimulation and teaching are needed to reach competency. This concept has been extended to other species like birds, such as the zebra finch nestling that must hear its father's song during a critical window of development in order to develop the song that is typical for that species.

Thus, while humans are highly skilled at language acquisition—with even precocious preschoolers speaking in complete, adult-like sentences, sometimes in multiple languages—language is hardwired in humans only in the sense that it emerges if and when the proper developmental supports are in place during the necessary phase of development. Moreover, given the preponderance of cases such as dyslexia, autism, and speech impediments, there are likely hundreds if not thousands of ways in which this developmental sequence can become disturbed, delayed, or altered, many of which trace to environmental and not just genetic processes.

And so it goes with offspring care and altruism, across species. For example, macaque females that are not treated well during their childhood also become insensitive mothers when they reach adulthood.[6] More recent and applicable experiments into the early development of rats have shown that the "passive" licking and grooming that dams provide to their newborns in the nest are influential for developing the pups' later behavior and physiology. For example, pups that receive more of this species-typical grooming stimulation from their mothers in the nest develop a stronger capacity to regulate affect, respond to stressors, and they develop different neural interactions between estrogen and oxytocin in the medial preoptic area (MPOA), the critical brain area in the hypothalamus for pup retrieval.[7]

Thus, I do consider and describe altruistic responses as instinctual at a motivational and motor-preparatory level. For example,
I refer to the attributes that lead to an altruistic response (e.g., a
vulnerable target in immediate need) as "releasers," just as ethologists described the instinct of greylag geese to retrieve their eggs
in the 1900s. The very word "instinct" or "instinctual" sends up
red flags because instinctual behavior is assumed to be relegated
to "infrahuman" animals (a term that used to be applied to nonhuman apes or monkeys by scientists who considered these animals "lower" on some imagined evolutionary hierarchy), above
which we have surely risen. This "humans are special" argument
is tied to the belief that our decisions (all of them and only ours)
reflect rational, explicit, deliberative cognitive processes—
certainly not urges or instincts. What are we, animals? Indeed,
we are. How can we make rational decisions if we are following
the base instincts of rats? How can we elegantly shift our decisions with the context, situation, individual, or mood if we follow instincts?

To solve this problem, we must leave simple stereotypes of evolutionary theory and the division between nonhuman and human
animals. We must examine the artful beauty of the brain itself,
so that we may appreciate how it is able by design to promote
(and prevent) giving in ways that were adaptive over a long period
of time. This artful design is even present in the tiny, Brazil nut–
sized brains of rodents. Far from simple, the mammalian central
nervous system evolved continuously for two hundred million
years, to solve problems that we all face as animals, such as how
to find food, obtain mates, and ensure the survival of offspring.
Even if humans did not have any additional cognitive capacities
over our rodent brethren, the mammalian brain would still foster motivated action in cases where we feel bonded to the other
and capable of helping but not scared, uncertain, or possessing

competing goals—when it is a good idea. As such, this book is less devoted to proving that humans are specifically born to be helpful and more to showing that an "instinct" to help can be predictable and occur under constrained circumstances that have been adaptive, owing to our evolutionary past.

As early as 1908, early social psychologist William McDougall similarly argued that "when we see, or hear of, the ill-treatment of any weak, defenceless creature (especially, of course, if the creature be a child) tender emotion and the protective impulse are aroused on its behalf. . . . The response is as direct and instantaneous as the mother's emotion at the cry of her child or her impulse to fly to its defence; and it is essentially the same process."[8] This "simple" but elegant neural design, which has existed for millennia, is surely augmented in humans by cortical processes that support our extensive learning, strategy, and ability to inhibit urges when they compete with our own longer-term goals. Even so, the altruistic urge has a lot in common with those fixed action pattern instincts that were described in the early days of ethology.

RETRIEVAL AS A FIXED
ACTION PATTERN

According to the altruistic response model, offspring retrieval represents a sort of "fixed action pattern," that can be released toward non-offspring, under conditions that mimic a helpless offspring in need. This is similar to the "misplaced parental care hypothesis" for avian cooperative breeding.[9] In an often-described fixed action pattern from Konrad Lorenz and Niko Tinbergen, a greylag goose will sit upon and retrieve her eggs when they roll out of the nest.[10] The geese do not retrieve only their own eggs but will also retrieve any other object that looks similar to her egg, under maternal conditions. As James Gould described it, "egg rolling behavior is striking: when an incubating goose notices an egg near the nest, its attention is suddenly riveted. It fixates on the egg, slowly rises, extends its neck over the egg, and with the bottom of its bill painstakingly rolls the egg back up into the nest."[11]

This egg retrieval behavior was considered an encapsulated and fixed "motor program," in the sense that once the act was released by the sight of the egg, the motion of the goose's neck drawing the egg back into the nest continued to completion even if the egg was removed by the experimenters. The action was also not highly specific to her own eggs, since geese would also retrieve other objects similar to eggs that were different sizes and colors: baseballs, rocks, and even beer cans and a small white animal skull. The action did involve an initial "poking" action after the neck protrusion to ensure that it was the right type of object (e.g., she would eat hardboiled eggs and reject squishy objects after poking), and they never recovered items with corners or points. Moreover, the geese would retrieve "supernormal" stimuli, such as the large eggs of other species, even more quickly than their own eggs.[12] As an extreme example, the geese even preferred to retrieve a volleyball over their own eggs.

Figure 4.1 depicts the images that Lorenz and Tinbergen captured during their studies and reported in their 1939 article. The first three panels depict first her noting, rolling back, and sitting upon a normal egg to protect it; the last image depicts her attempting to retrieve an oversized fabricated Easter egg, which she tried to roll back but could not complete because of the size, leaving her looking "in embarrassment." Because geese need a way to ensure that their eggs stayed warm and safe in the nest, the brain evolved a fixed action pattern that is encoded

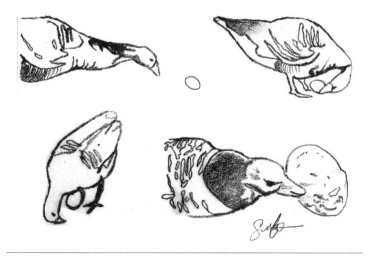

FIGURE 4.1 Drawn depiction of the sequence of movements when greylag geese retrieve their eggs back into to the nest, including both normal and supernormal stimuli, which the geese also retrieve.

Redrawing by Sarah N. Stansfield, CC-BY-SA-4.0, based on information in Konrad Lorenz and Nikolaas Tinbergen, "Taxis und Instinkhandlung in der Eirollbewegung der Graugans [Directed and Instinctive Behavior in the Egg Rolling Movements of the Gray Goose]," *Zeitschrift für Tierpsychologie* 2 (1938): 1–29.

with this response to any object nearby or rolling away that is egglike in shape (i.e., smooth, rounded, convex). The objects are called "releasers" of the fixed action pattern, based on features that are referred to as "sign stimuli." I mention these details because I consider features of the victim and situation to resemble these sign stimuli, which in turn "release" the rescue response in the observer, in much the same way that geese pull an egg or rounded object toward the nest. Thus, I sometimes refer to these cues of need in victims as "releasing" a rescue response in observers—like a preprogrammed and routinized motor act that stands at the ready, to be issued forth under the right conditions.

This fixed action pattern of the geese likely evolved because of the enormous selection pressure to reliably retrieve and protect eggs that have fallen out of the nest. The fact that this mechanism can produce odd, accidental retrievals of beer cans is less problematic in nature, where there are fewer objects to accidentally release the sequence. A clear exception is in the case of "brood parasitism," in which an animal like the "screaming cowbird" of South America places its large egg into the nests of other species to be fostered by the new mother.[13] The larger, louder cowbird offspring are "supernormal" releasers of the host bird's incubation and feeding response, which can result in the host giving food to the intruders even before their own related offspring. (Some say the cowbird will even peck and kill the host's eggs if they try to remove it, in which case, even recognizing the eggs or chicks as unrelated would not help.) There are ways to keep this retrieval response in check, such as with a built-in capacity to detect and remove unwanted items from the nest, like the beer can that was accidentally retrieved previously. Egg retrieval is also kept in check by the fact that it is released only between the phases of incubating and hatching, which prevents her from accommodating others' eggs or strange objects "any old time."

There are many similarities between this early ethological discovery in geese and the rodent pup retrieval from which we began. Both behaviors are literally retrievals by mothers of offspring, since mothers in both cases grab back an offspring that has become separated from the safety and warmth of the nest. Both retrievals are observed in females during parturition and evolved to protect their related and helpless offspring. Both retrievals also involve a highly motivating act that is most likely to occur when the animal needs it most. Pup retrieval may not be a literal fixed-action pattern in the sense that it does not appear to be released as a complete motor act from the moment of detection through the final motor commands, but there are signs that it is at least somewhat fixed. For example, when early neurobiologists tried to determine where in the brain the rodents' offspring retrieval response was encoded, they started by making large lesions to narrow down the critical brain region. When they lesioned the anterior cingulate cortex (ACC), which is associated with detecting problems in the internal body or external environment to facilitate a response, sometimes a dam would do odd things like trying to retrieve her own tail back into the nest. This revelation, another quick note added by researchers in an old article that would probably not make it into a modern manuscript, suggests a fairly fixed motor plan to retrieve that is somehow refined by the cingulate under normal conditions.[14] Denoting this commonality between fixed action patterns in ethology and rodent caregiving, Burton Slotnick applied the fixed action pattern of Tinbergen's stickleback fish to explain maternal care sequences in rodents.[15]

Despite being instinctual, even fixed-action patterns are not considered by modern biologists to be encapsulated, innate, unalterable, or uncontrollable. Rather, these actions are considered to be spontaneous, stereotyped behaviors that are (1) hard to control once enacted, (2) expressed by all typically developing members of the species, and (3) subject to contextual and epigenetic effects.[16] For example, when extending fixed action patterns to rodent caregiving, Slotnick specified that the care sequences would not be as fixed and as hierarchical as in fish, and his proposal included a flexible response organized by frontal, cingulate, and septal areas.[17] Mammalian neural systems are inherently goal-directed and context-sensitive. Therefore, even behaviors assumed to be "innate" are not truly inflexible or noncognitive: they simply reflect an implicit decision that is highly motivated and that maximizes one's key goals, while still reflecting the one's own genes, developmental history, and current context.

HARDWIRED TO ACT?

The term "hardwired" is most often used when people refer to the behavior of an animal that they deem simple, or when non-scientists discuss human behaviors that seemingly occur without effort or learning. In point of fact, scientists almost never use the term "hardwired" (except disparagingly, perhaps) because it is virtually always misleading. The term conceals the fact that even multicellular organisms have genetically based, neurophysiological mechanisms that produce divergent responses from the organism's early and current environment. There is simply no strict division in most of biology between nature and nurture. Even amoebas exhibit context-sensitive altruism and individual differences. For example, asexual free-living cells form a mass of cells into a slug when food is scarce that can "reach" for a new food source. This slug includes "cheater" cells that compete to be in the spores of the fruiting body that make it into the new and hopefully richer environment, while other cells altruistically form the sterile stalk that is left behind in the poor environment.[18]

On the one hand, the general premise of the altruistic response model is simple: human altruism reflects our heritage as care-giving mammals that evolved the propensity to care for vulnerable targets that we can help. On the other hand, there are many caveats and complexities to this perspective that must be appreciated to avoid oversimplifying this already simple theory, so as to make it accurate and not just pithy.

Theories that seem simple—like one rooted in homologous brains or instinctual goodness—make for great straw men for scientists who make hay by complicating others' phenomena. For example, my perception-action theory of empathy with Frans de Waal was interpreted to mean that people automatically mimic others' emotions and feelings.[19] Full stop. A slew of articles followed to criticize the theory because, of course, people don't just walk around mimicking every expression they observe, and it is fairly easy to demonstrate changes in empathy by context, attention, competing goals, or top-down cognitive processes—features that were already explicit in the original theory but did not survive the tendency to read for and remember gist only (or, less generously, to misconstrue others' theories to topple or outshine them).

I am not Richard Dawkins in most ways. But I sympathize with his famous description of being misrepresented when people overlooked the complexities of his seemingly simple theory: that genes are "selfish." After writing *The Selfish Gene*, Dawkins had to explain, time and again, that his gene-centered theory does not mean that people *themselves* are only selfish.[20] Dawkins's apocryphal witticism on this point must be true since he recently integrated it into the introduction of his thirtieth anniversary edition, where he wrote about regretting his book title, claiming that "many critics, especially vociferous ones learned in philosophy as I have discovered, prefer to read a book by title only. No doubt this works well enough for The Tale of Benjamin Bunny or The Decline and Fall of the Roman Empire, but I can readily see that 'The Selfish Gene' on its own, without the large footnote of the book itself, might give an inadequate impression of its contents."

Because people have such a strong tendency to oversimplify theories, I focus here on addressing common misunderstandings over presenting the basic science of offspring care, the latter of which was the focus of the original academic article.[21]

At the level of the species, the behavior of William E. Wilsoncroft's dams may suggest that offspring retrieval is hardwired. But important caveats of what it means to be hardwired must be appreciated to contextualize this statement, even if I am fine with that characterization of pup retrieval. The mechanism for pup retrieval already relies upon a complex mix of genes, hormonal releasers, and situational factors that make even this hardwired behavior fairly flexible and sensible. This description of how things are hardwired into the nervous system is more complex than a simple or encapsulated fixed-action pattern and requires that we understand how nature naturally integrates with nurture, through elegant designs in the nervous system.

Decades of research on offspring care since early studies like Wilsoncroft's have demonstrated that offspring retrieval only occurs under conditions that make adaptive sense. For example, it would be bad if all unmated rats were highly sensitive to pups' needs if they naturally encounter unfamiliar and unrelated neonates in their daily routine. Moreover, the common but powerful tendency by animals to avoid novelty helps them to avoid myriad dangers, from strange food to strange members of their species, predators, and wide-open spaces. Thus, the avoidance-approach dichotomy that is built into the nervous system allows rats to shift from avoiding the needs of pups to becoming diligent caregivers. Hormones such as estrogen and progesterone shift over the course of dams' pregnancy, with particularly robust changes at the time of parturition. These hormonal shifts actually change the dams' brains, rendering pups highly salient and rewarding stimuli that dams seek to retrieve and huddle. Scientists have demonstrated this process in myriad ways, such as measuring how these hormones change naturally during pregnancy and parturition, artificially removing or blocking their effectiveness to observe the impacts on retrieval, or administering hormones to unmated virgin females or males to potentiate a retrieval response.[22]

Thus, even though rodent offspring retrieval is hardwired in the sense that all typically developing dams do it under normal circumstances, it is hardwired only through a natural cascade of events that correspond to the need to gestate, deliver, and care for one's own pups. The process is undergirded by a complex interaction between genes, perinatal sex hormones, neurotransmitters, and their effects on the brain, all of which can be altered or impaired. Even in the original Wilsoncroft study, there were large individual differences across the dams, demonstrating how the response is altered by many converging factors rather than a fait accompli. As with the greylag goose, the dam's motivation to retrieve is limited to the first weeks after pups are born and flags once more habitual care takes over and the neonates can fend for themselves. For example, just after birth, dams prefer to access to her own pups over cocaine, but this noble preference shifts back toward the stimulating drug after a few weeks.[23] Thus, pup retrieval by dams, just like the egg retrieval of greylag geese, is hardwired, but in a way that is sensibly released by internal and external cues of a helpless offspring, during the critical neonatal period.

HELPING STRANGERS
IS NOT AN "ERROR"

People often assume that if our altruistic responses originated in offspring care, then extending them to human strangers must be an error or mistake, even one that should be eliminated. Just as we assumed that greylag geese made a mistake when they retrieved a beer can, we might assume that it is a mistake to leap into a subway track to retrieve a strange young man from an oncoming car. In short: if evolution's goal is to preserve and promote our own successful genes, we should not be saving strangers.

Further still, if altruistic responses are indeed mistakes, then they could be eradicated once evolution has had enough time to weed out the annoying glitch. In ten thousand or a hundred thousand years, people who were unwitting enough to rescue strangers from capsized boats or donate to orphaned children far away should be outperformed by the more discriminating among us who help only related individuals or those who will repay our kindness or otherwise benefit us. The belief that our empathy-based aid should be eradicated is not actually a straw man, inasmuch as modern authors such as Paul Bloom argue "against empathy": that we should only make decisions to help that logically maximize the greater good while suppressing our all too emotional, damaging, pathetic, and misdirected sympathy for those less fortunate.[24] Perhaps the long arm of evolution will eventually eradicate this foolhardy generosity, but by then we will all be dead, and I will not have to hear you say, "I told you so!" Even so, there are many logical reasons to consider this extended care to strangers as something other than an error or temporary glitch:

1. At a practical level, it is difficult to adjust a mechanism that protects helpless offspring just to avoid a few accidental or even intentional extensions to strangers. The fitness costs of inadvertently impeding the primary goal of protecting offspring would be much worse than the benefits recouped by preventing occasional extensions to others.

2. The avoidance-approach opponency that is baked into the neural circuit *already* balances our own and others' needs by linking aid to caregiving, which is released only by vulnerable

victims in clear need that we can help, when we are not overly fearful or uncertain.

3. The offspring care mechanism may already be refined to prevent disadvantageous responses. For example, Sarah Hrdy postulated that the caregiving instinct was modified throughout primate and hominid evolution to allow for more calculating and controlled forms of sympathy.[25] (Hominids did develop more control over behavior in general, through expanded executive processes, but I am more compelled by how instincts themselves are implemented to prevent "erroneous" aid.)

4. Sometimes what looks like an error in our wiring just reflects unavoidable individual differences. A "normal distribution" of individual differences naturally occurs whenever a behavior is instantiated by multiple underlying and interacting genes, with some individuals on the lower end of the spectrum and others on the high end, with most in the middle. Applied to altruism, if an unfamiliar toddler started falling as the city bus accelerated, one rider might lunge to save him while another sat idly by laughing, but most riders would be concerned and want to help even if they remained seated unless the child were near or in great danger and they knew they could help. Individual differences in observers (e.g., sensitivity to novelty, infants, risk, or to perceiving need) are described in a later chapter. These biases naturally produce a distribution of responses across individuals that is usually unproblematic and is unavoidable when behaviors rely upon multiple genes and impact of one's environment.

5. The majority of human altruistic responses involve small, calculated costs, like donating a few dollars or minutes of your time. Thus, even though you might observe what seems like erroneous types of altruism, such as psychopathy on the one hand versus devoting your life to the poor on the other,

these extremes signal a system that is largely kept in check
and is generally adaptive.

6. Beyond the small or rare costs that most of us suffer from being
    altruistic, there are even fitness *benefits* to helping if individuals
    in our social group who share genes or could help us later. Our
    beneficence is observed and can be reciprocated (to us or our
    kin) from the initial victim (direct reciprocity) or from anyone
    who appreciates our act (indirect reciprocity). The group as a
    whole also benefits from a cooperative spirit, which relies at
    least in part on our instinct to care. Altruistic responses also
    ameliorate the negative impact of prolonged stress or distress
    on health, group harmony, and predation risk.[26] Moreover, it
    feels good to help, through releases of dopamine and oxyto-
    cin.[27] Thus, the tendency to give does not come only with a cost
    but also with real benefits to ourselves and those around us.

For all of these reasons, I do not think that altruistic responding should be considered an error or one that should be or will
be eradicated by evolution. The instinct is necessary for one's own
reproductive success, which makes it difficult to constrain, it is
balanced by a mechanism that is only released in specific (largely
adaptive) circumstances, and it has already been refined somewhat to permit strategic and controlled giving. Moreover, only a
minority of individuals, on the tail ends of the distribution, are
predisposed to emit problematic responses, whereas most of us
only give small, low-cost gifts that come with benefits from
inclusive fitness, reciprocity, group cohesion, and improved
affect and health. As such, altruistic responding is by all accounts
adaptive while accommodating existing ultimate-level views of
altruism that focus on later benefits through inclusive fitness,
reciprocity, and cooperation (see chapter 9).[28]

SYSTEMATIC ERRORS IN OUR
EVOLVED SYSTEM

Other chapters explain how the brain evolved to implicitly integrate cues into one holistic interpretation, such that our best guess is fairly accurate, presuming we attended to the relevant cues. But this system makes us susceptible to systematic biases that are embedded in the design that is otherwise and still adaptive. Thus, things do not *always* work out for us, but the problems that do arise are predictable from the way the mechanism works, giving us the chance to avoid them. Biases in the way our system work particularly produce errors through misperceptions of the "sign stimuli"—such as whether the victim really is vulnerable, helpless, in urgent need or we think we will succeed.

Research has shown a wide variety of ways in which our perception and behavior can be altered by our distant and immediate past, in ways that change genes, hormones, and behavior. As such, any one person can enter into a situation with a systematic tendency to underpredict or overpredict a particular state or outcome, representing true errors. These errors do not mean that altruism writ large is an error, but they can be objectively classified as errors in the sense that they either fly in the face of objective risk or probability or do not fit with one's own goals, values, or plans.

Errors from the Link Between Distress and Need

As an example of a learned bias, an observer who was abused as a child may erroneously assume that whenever they make a mistake or upset someone they will be physically or verbally assaulted.

This observer carries their learned and more or less adaptive expectation of punishment from early childhood into adult life, even if they are later surrounded by kind and accommodating people. As a result, an abuse victim may fear angering someone and try relentlessly to avoid mistakes, to keep others happy, and to escape anyone who may be upset. For example, our first beloved pet terrier, Kermit, was rescued from the streets of San Jose, California. He had clearly gone through some tough times before he was rescued, because for years after his adoption if anyone said "no!" or raised their voice—even from excitement and not anger, even on TV—he would cower in fear, lowering his head and hiding his tail between his legs as he slowly backed away, awaiting whatever terrible outcome he had come to expect from life before adoption. This hypervigilance to punishment lasted for years before Kermit eventually learned the ways of his loving caretakers, but it still peeked out at times. Responses that are genuinely adaptive in one's early environment can make life difficult later, such as when adults try to trust their kind romantic partner after having suffered abuse.[29]

Applied to altruism, someone without positive early childhood experiences may be reticent as an adult to show weakness or ask for help, even in a genuinely trying situation. This learned reticence may be adaptive in the face of caregivers who are chronically annoyed or rejecting, but later in life it may backfire around friends or family who genuinely want to help. People from this difficult background may never display their need clearly enough to show their suffering, making it hard for those around them to respond appropriately. Reticent people may even reject others' distress, which they learned by observing their own irascible caregiver.

In an experiment that we performed at the University of Iowa Hospitals and Clinics, we videotaped interviews with patients suffering from a variety of terminal and serious illnesses, including cancer, hepatitis, kidney failure, and heart disease.[30] The most noticeable thing about these interviews was that even though all patients had serious health problems, they expressed themselves very differently. The "sanguine" patients were happy and upbeat, even working to make the interviewer feel comfortable by making jokes or smiling. The "reticent" patients were quiet and reserved and avoided sharing their problems. The "wistful" patients were sad and thoughtful without being overly emotional. The "distraught" patients were emotional and cried throughout the interview, reflecting on their illness and love of family.

Consistently, people who watched these interviews agreed that distraught patients needed the most support, but even so they sometimes wanted to avoid these patients and preferred to help the sanguine ones who seemed happy and less in need. The most reticent patient did not want to talk about his problems at all, gave one- or two-word answers, avoided the camera, and looked uncomfortable throughout. His response did not seem that unusual for an older farmer from the Midwest, but our observers did not perceive his need because he didn't talk about it or display distress, and they subsequently offered him the least empathy and money, even though his burden was just as great as for the other patients. You could perhaps read through the farmer's reticence and see the pain "behind his eyes"; indeed, our most empathic observers felt more empathy for him than the average person, who generally did not detect his too-well-veiled need, limiting their desire to help. Thus, because we infer need from others' distress, we can fail to identify genuine need, even high levels of need, if we are trained to hide or ignore such vulnerability.

The requirement for distress to indicate need can also cause people to mistakenly assume that there is or is not an urgent medical emergency at hand. For example, many of us have raced across the house to check on a family member screaming in pain, only to discover that they stubbed their toe or hit their funny bone on a bed corner—not an emergency. Conversely, some genuine medical emergencies do not seem urgent because they lack the "releasing" cues or sign stimuli of need. For example, the female heart attack is less often accompanied by a sharp intense pain near the heart, which people consider the telltale sign of a heart attack. Added to this, medical professionals are biased against perceiving, believing, and treating pain in women, compared to men. As a result, many females have had heart attacks that made them feel weak, sweaty, or flulike and so are either not treated for a heart attack or die on their beds at home, wondering why they feel so sick and hoping it will just pass. As evidence, the amount of time that passes between first feeling symptoms to seeing a medical professional is 34 percent longer in women than men, and once they arrive at a hospital women wait 23 percent longer for reperfusion therapy than men.[31]

In another common and difficult case, people can have a stroke that does not produce signals of distress or injury and so are not treated for the life-threatening injury. For example, even young, healthy people can have a stroke after sustaining a sports-related head injury, for example, from heading or being hit by a ball, colliding with another player, or skiing into a tree. These injuries may not seem life-threatening if the victim is not bleeding or screaming unless he gets a severe headache from the buildup of blood. When the stroke finally manifests, after a delayed period during which fluid built up in the brain, the victim can appear "odd," but not in a way that warrants a trip to the emergency room. For example, a stroke or traumatic brain injury can lead to strange, nonsensical utterances or numbness or blindness on one side of the body. Observers sometimes initially laugh because the victim's behavior seems silly and does not release our response to need. For example, the actress Natasha Richardson was at a ski lesson in Quebec when she fell and hit her head. At the time she did not show notable signs of injury, and the paramedics were turned away. Later, at her hotel, another ambulance was called because she "was not feeling good." She died two days later from an epidural hematoma. The ambulance manager was quoted as saying: "When you have a head trauma you can bleed. It can deteriorate in a few hours or a few days. People don't realize it can be very serious. We warn them they can die and sometimes they start to laugh. They don't take it seriously."[32]

Another unfortunate and common tragedy occurs when people drown silently, sometimes even when they can swim or are surrounded by potential witnesses. More than a minor issue, in the past few decades, among accidental deaths, drowning was the second leading cause of death in children and the leading cause in younger children one to two years old.[33] It is difficult for nearby bathers and even lifeguards to notice when someone stays underwater too long because there is no thrashing or screaming such as you expect from the movies when Jaws attacks. If someone simply does not come back up, there is no cue to take notice. A lack of information is not well regarded as a form of information by the brain, except if you have a strong prior expectation for an outcome that is withheld, like the rat that hears the tone but does not get the juice or the dinner guest who waits for a dessert that never arrives.

These unusual of emergencies—female heart attacks, strokes, drownings—are not only terrible because of their drastic and lasting effects, but they are also difficult to redress because our brains evolved to so tightly link need to the sign stimuli and releasers of neonatal distress. It is important to understand this evolutionary heritage—with its attendant benefits for altruism or heroism and its potential to foment tragedy—when designing public health messages and appeals for aid.

Individual Differences in Risk Perception

One of the most commonly studied biases in decision making is the tendency to be risk averse, documented in the popular book by Daniel Kahneman, *Thinking, Fast and Slow*.[34] This book describes Kahneman's decades of research in behavioral economics, which demonstrates this bias and many others. According to the research, people differ in the degree that they avoid versus seek risk, but animals as a whole are biased to be more risk-averse than not because this adaptively prevents threats to survival in the face of uncertainty. Thus, if a monkey or child is faced with two foods to eat, one that is novel and one that is familiar, each is more likely to eat the familiar one. If they do eat the novel one, they do so more slowly, or in smaller quantities, experimenting first to avoid being poisoned.[35] Being cautious has served us well over time.

In the context of altruism, bystander apathy is a good example of risk aversion.[36] People may perceive the victim's distress and understand that it is an emergency yet remain uncertain about what exactly what the problem is, how to help, and the possible consequences. This uncertainty biases them to avoid responding. As the old adage goes, "better safe than sorry," especially when the victim is an unrelated stranger who is not interdependent with you, and other onlookers seem more responsible or certain.

Displaying our quick predictions, figure 4.2 shows a woman who looks either happy or angry.[37] On the extremes to the far left and far right, it is easy to tell how she feels. From about 0 to 30 she looks happy, and from about 70 to 100 she looks angry.

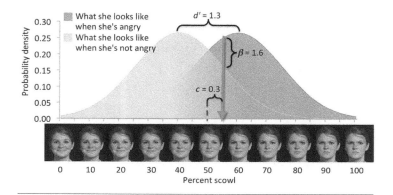

FIGURE 4.2 Figure from a study that applies signal detection theory to demonstrate the continuum of facial expressions, which produce more variable impressions on people in intermediate cases.

From Spencer K. Lynn, Jolie B. Wormwood, Lisa F. Barrett, and Karen S. Quigley, "Decision Making from Economic and Signal Detection Perspectives: Development of an Integrated Framework," *Frontiers in Psychology* (July 8, 2015), https://doi.org/10 .3389/fpsyg.2015.00952.

But what about in the middle? According to signal detection theory,[38] our uncertainty between classifying something as one thing or another is represented in the overlapping region between the two peaks. Some are biased to see her as more happy than angry and vice versa in the gray area between the faces upon which people usually agree. A person with a strong bias (e.g., from anxiety or abuse) may perceive anger at a low percentage of scowl, which most of us would not see as angry. Generally, the brain seems wired to prefer avoiding "misses" (thinking the person was happy when she was really mad) than "false alarms" (thinking the person was mad when she was happy), which is further exacerbated under fatigue, time pressure, stress, or when you expect the worst for other reasons, such as those mentioned earlier.

These risk biases are generally adaptive, but they sometimes promote appalling behaviors that are hard to override. For example, an officer of the law may feel stressed and scared in a quickly changing situation in a dangerous neighborhood that he associates with violence. His association reflects both learned and inaccurate stereotypes about people of color and poverty, alongside a natural tendency to overperceive threat to avoid missing a detection. To address this problem, negative stereotypes about people of color or in poverty need to change, and there must be a clear, known, and scary punishment to misperceiving risk where it was not present. People can avoid acting on a bias when they expect a repercussion—because of the way brains naturally integrate past experiences with current cues. For example, even monkeys and human toddlers automatically and without deliberation redirect anger after being clobbered by a dominant bully onto a weaker subordinate individual, the latter of which will be more likely put up with it than the scary peer who will hit back.[39] Men beat their wives at home and aggress against women more than men in the workplace simply because they predict—even implicitly—that they can get away with it.[40] This does not require conscious consideration; people can learn through experience to associate objects, people, and situations with their attendant risks and rewards, which in turn bias decisions, even unconsciously.[41] Thus, to address genuine errors in our biased nervous system, we must change how we implicitly construe others and rebalance incentives to prevent even "sensible" seeming instincts from causing behaviors that are truly unacceptable.

Local Priming

We talked earlier about biases that are either built into the nervous system by design or learned from one's early developmental period. In addition, people can be affected temporarily by immediately preceding events or "primes" that biases their perception, even if that bias is not characteristic. As in a famous adage, "You are always fighting the last war," people with a particularly bad (or good) recent experience subsequently overpredict the same outcome, leading to errors. For example, you could meet a perfectly great future mate that you reject out of hand because your last relationship left you tired and hopeless, or because the person randomly said something that reminded you of a despised former lover, say, mentioning their cat or their mother. The monkey that was just beaten by a dominant individual enters the next situation primed with anger or fear, causing it to displace the aggression onto an unwitting subordinate that happens to be nearby.[42] These responses need not be planned. Note that even this fairly reflexive displacement down the status hierarchy is sensitive to context—after all, neither monkey nor human uses the opportunity to lash out at the leader of the troupe. As a positive example, a local, "context-setting" bias—a phenomenon that is even observed in *Drosophila* fruit flies—can lead to good outcomes, such as when people "pay it forward" and are nicer to other people after receiving a random act of kindness from a stranger.[43]

SUMMARY

As the title of this book suggests, it is easy to remember that altruism is an urge or instinct that derives from the need to care for helpless offspring. However, many will skip over or forget the details and perhaps only vaguely recall that this book "has something to do with altruism being an urge, like caring for babies." In order for this gloss to make sense, you also have to remember that the specifications of the model clarify exactly when this urge occurs, which reflects a complex mixture of genes, early environment, individual differences, and the situation. The altruistic response model uniquely emphasizes that even as a "fixed action pattern," the altruistic response will not occur in just any person or situation but is instead "released" by "sign stimuli" that were relevant to the context of caring for offspring: when the victim is neotenous, vulnerable, helpless, and in immediate need of aid that the observer can provide. Each of these features, and our propensity to detect them, are altered by our personal experience and expertise, reflecting the natural intertwining of nature and nurture in our nervous system and behavior. By accommodating these more nuanced aspects of the model, we can move beyond a generic belief that people evolved to be naturally helpful and understand when we will and will not act.

5

THE NEURAL BASES
OF ALTRUISM

This chapter reviews key aspects of the neural and hormonal bases for altruistic responding. Extensive descriptions and supporting evidence are provided in my academic paper on the altruistic response model.[1] To help people appreciate the role of the offspring-care system in caregiving and altruism, I focus on explaining a few key attributes such as the neural opponency between approaching and avoiding victims, the role for inherent rewards and oxytocin, and when this neural system participants in human altruism.

KEY FEATURES OF THE NEURAL
CIRCUIT FOR OFFSPRING CARE THAT
EXPLAIN ALTRUISM

The prior chapters described a neural circuit in rodents that supports passive and active offspring care, in which individuals shift from an initial avoidance of strange, novel pups to actively approaching them when induced into a parental state (figure 5.1). This neural opposition that it built into the brain circuitry is fundamental to our ability to understand human altruism, which is

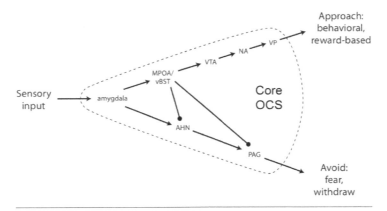

FIGURE 5.1 A depiction of the neural circuits that support offspring care, from research on rodent pup retrieval, referred to collectively as the offspring care system.

Stephanie D. Preston, "The Origins of Altruism in Offspring Care," *Psychological Bulletin* 139, no. 6 (2013): 1305–41, https://doi.org/10.1037/a0031755, published by APA and reprinted with permission, License Number 5085370791674 from 6/10/2021.

similarly characterized by both embarrassing apathy and urges to respond.

The Opponency Between Avoiding and Approaching Others

The opponency between avoiding versus approaching rodent pups is part of a more general way that we understand neural processes, which use opposing states to balance divergent behaviors. This concept was a focus of Theodore Christian Schneirla,[2] who was a 1925 graduate of the University of Michigan before he became a professor at New York University and a curator at the American Museum of Natural History. Schneirla's early work explored army ant raids in Panama, but his concept of opposition has since been applied to a wide variety of phenomena in psychology, including personality, psychopathology, brain lateralization, and collective group behavior. Schneirla mentored the neuroethologist Jay Rosenblatt, the longtime director of the Institute of Animal Behavior at Rutgers University–Newark, who pioneered the application of this concept of opponency to offspring care in animal models. For example he demonstrated that nonparents will also retrieve pups if they have time to habituate to them or are given the necessary neurohormones associated with pregnancy.[3] His work has continued to the present through his former mentees Michael Numan, Alison Fleming, and Joe Lonstein.

In the animal model of offspring care, the perception of a pup activates the amygdala, which participates in both the avoidance and the approach route of the circuit. In animals that have not yet mated and are not caring for pups, the avoidance circuit proceeds from the amygdala to the anterior hypothalamus (AHN) and then on to the periaqueductal gray (PAG) of the brainstem. The PAG is situated at the bottom of the brain near the spinal column, and its neurons go on to alter processes in the body, such as increasing arousal and promoting behaviors that avoid the novel pups. Avoidance is considered the "default" state because it is the nonparental state from which most rodents begin.

In a rodent that has prepared for parenting, this default avoidance system is instead *inhibited* by the amygdala, which then instead projects to regions of the hypothalamus (with names that seem unnecessarily complicated: the medial preoptic area (MPOA) of the hypothalamus and the ventral portion of the bed nucleus of the stria terminalis (vBST)). From these ancient hypothalamic areas, deep in the middle of the brain, the ventral striatum, an area that is dense with receptors for the reward-related neurotransmitter dopamine, is activated next, which motivates animals to actively approach the pups.

Once the pup is retrieved through the approach circuit, and is safely contained in the nest, the ensuing close contact between the adult and the infant provides additional rewards to both through the opiate system, which enhance the rewarding signals from dopamine in the ventral striatum (e.g., the nucleus accumbens, NAcc), further motivating future approaches. There are also connections that involve glutamate between the prefrontal cortex (PFC), hippocampus, and NAcc, which additionally boost the growing positive association between the pups and the rewarding signals emanating from the NAcc via the MPOA. Neurohormones that are essential for offspring care across species, such as oxytocin and vasopressin, also support dams' motivation to approach and foster a long-term bond with pups and a memory for their identity in the NAcc.[4] These processes combine to ensure that new mother dams are highly motivated to attend to and care for pups—an effortful but adaptive and critical process for survival of both pup and dam.

Of note, most regions in this neural circuit do not only handle offspring care, except perhaps the MPOA. For example, the nucleus accumbens and its rewarding dopamine neurotransmitter are involved whenever an organism is motivated toward an attractive, rewarded, or important target of any type—including consumable items like food and drugs, as well as rewards that cannot be literally consumed, like pups, money, or fancy clothes.[5] Offspring care might not have even been the initial target of this reward system, inasmuch as early mammals would have had to acquire food and mates before they cared for offspring for an extended period, sometime in the late Triassic period. This domain-general property of the neural system is important to realize because when people hear terms like "caregiving system" or "offspring care circuit," they infer that these areas are *for* this behavior and this behavior alone. As a general principle, brain regions rarely support behavior in just one domain. As I tell my students repeatedly, "There is no altruism area!" Of course, sections of cortex prefer certain types of information, such as faces or houses or pups that need retrieving; however, these brain areas still participate in a larger system that is activated by any similar information or stimuli.

The Experienced and Neural Rewards of Helping

Remember back to William E. Wilsoncroft's assiduous dams, which retrieved unrelated pups for literally hours until the exhausted experimenters quit trying to find their breaking point.[6] In that study, the females had to press a bar for the pups, which caused the pups to fall down the chute into the chamber. By design, there was no reason for the dams to press the bar in any phase of the experiment. The females could have just sat in the nest and relaxed. This is particularly true once the experimenters removed the initial unconditioned rewards of food or related pups. Rats in most conditioning experiments do stop pressing the bar after a series of trials or blocks after the food rewards are eliminated.[7] Why did the dams continue to press the bar?

There are multiple conceivable explanations for this odd fact. Maybe the dams couldn't unlearn the strong association between bar presses and food or offspring, or kept pressing in the hopes that eventually more food or related pups would appear, particularly if a dam received only the first six pups and still had more that did not shoot down the tube. These explanations are unlikely, however, because of the countless demonstrations of rats habituating after the rewards are withdrawn and their known ability to recognize their own pups.[8] Thus, if the dam were just waiting for more of her own, there was no reason to carry them back to the nest once she noted that subsequent ones were unfamiliar.

The fact that dams pressed the bar without any traditional reward, over and over again for hours, signaled to the experimenters that the dams were *rewarded by the arrival of pups themselves*, despite the fact that they were unrelated. In Skinnerian terms, the dams' bar pressing shows that they were so motivated by the pups that they were willing to work to obtain them—just as we would work for any other reward, such as food, water, alcohol, cocaine, money, or even praise. Contact with the pups was experienced as similarly pleasurable and destressing for dam and pup alike, which only further reinforced the motivation to retrieve, yet again. In this way, other individuals are generally rewarding and compelling stimuli *to which* we are driven to interact and *from which* we receive affective and physiological benefits.

These rewarding processes in the NAcc that involve dopamine have been shown to participants any time an individual "wants" an item that was rewarding in the past.[9] For example, dopamine levels change when a new dam prefers the cage associated with pups over an empty one, or a cage that contains another reward, such as cocaine. Moreover, if you remove dopamine from the ventral striatum, retrieval behavior declines; however, it resumes if the dams are deprived of pups beforehand.[10] Thus, just as food looks and tastes better when we are very hungry ("Hunger is the best pickle," said Benjamin Franklin), dams are even more motivated to seek pups when they have been deprived of their comforting contact. Unlike the MPOA, which is required for a retrieval, the NAcc is not *essential* for a retrieval. If the NAcc is damaged, dams can still prefer, nurse, build a nest, and press the bar for pups.[11] Lesions of the outer shell region of the NAcc do disrupt maternal behavior and pup retrieval, but only after a few trials or a day and not immediately, which we infer to mean that the NAcc is not truly necessary for the act of retrieval itself, but rather supports the continuation of the act under normal circumstances.

The outer shell region of the NAcc and opioids are associated with the way we "like" consumable rewards, as opposed to "wanting" them (as wanting is associated with the core of the NAcc and dopamine).[12] For example, if you increase opioids in the brain, passive and active maternal care also increase across species. Conversely, if you decrease opiates, the protection, retrieval, and grooming of pups decreases. The "focused preoccupation" of primate mothers with their infants is also eliminated without opioids.[13] Thus, the MPOA is *necessary* for pup retrieval, whereas dopamine and opiates in the NAcc are needed during the initial retrievals to ensure that pups become rewarding and motivating, which then facilitates the behavior until a habit is formed.

Applying this to human altruism, people can implicitly predict when it will feel good to help someone, which promotes care at times when it would be sensible, such as when a close other needs help or when helping would assuage our distress. Because this reward prediction is handled by these ancient neural circuits, the urge to help can occur in the absence of any conscious awareness or prediction of the reward of helping. This makes it all the more unfair that we should discount an altruistic act that came with a reward for helping, when we are not even necessarily aware of this future reward at the time of the offer. Sometimes people are clearly aware of the benefits of helping, but this type of strategic helping involves forms of cognitive and neural processing are not needed for an altruistic urge, even if they can complement the urge, as described in what follows.

Oxytocin Critically Reduces Avoidance and Increases Passive Care

Another well-studied attribute of this system is the role of the neuropeptide hormone oxytocin, which is critical for giving birth to, bonding with, and caring for neonates across species. In rodents, oxytocin reduces the natural avoidance of pups and promotes passive maternal behaviors like crouching, kyphosis, licking, and nursing.[14] In mice, maternal behaviors are severely impaired if the oxytocin gene (fosB gene) is removed.[15] The mice still approach, lick, and crouch over unrelated pups without this oxytocin, but they are less likely to pick them up or move them to safety.[16] Similarly, a lesion to the paraventricular nucleus (PVN), a brain area that is rich in oxytocin, causes dams to avoid and sometimes even cannibalize pups.[17] Pup retrieval can also be impaired by preventing oxytocin or vasopressin to act in the MPOA—that critical region in the hypothalamus for retrieving pups.[18] Multiple regions of the offspring-care system contain oxytocin receptors, including the VTA, MPOA, and NAcc.[19] Both oxytocin and dopamine act interact in the VTA and NAcc to promote responding. For example, after infusing oxytocin into the brain of a rodent, the mesolimbocortical dopamine system is activated.[20] Thus, significant evidence confirms a role for oxytocin in promoting the care of offspring.

EVIDENCE FOR A RELATED NEURAL SYSTEM FOR ALTRUISM IN HUMANS

From the evidence in animal models of offspring care in mice and rats—and other caregiving mammals like sheep and monkeys—it seems that oxytocin, like dopamine, is not *essential* for the ability to perform a retrieval but encourages retrieval by making females less anxious about approaching pups while facilitating the bond with and memory for the pups.[21] Applied to altruistic responding, oxytocin should reduce our avoidance of others in social situations so that we feel comfortable enough to approach them and to facilitate our bond with close social partners that we then continue to seek.

Similar Caregiving Processes in Humans

The mechanisms that support offspring retrieval in rodents are similar to those found in caregiving humans. Across mammals, newborns are attractive and pleasurable to cuddle with—to the point that monkeys sometimes "kidnap" others' babies so that they may hold them as their own.[22] Human grandparents can insist on holding their infant grandchildren, even if the mother protests that the infant should be sleeping in a crib, sitting in a high chair, or belted into a car seat. Perfect strangers try to touch babies at the grocery store and even the stomachs of pregnant women, to the horror of some. Sometimes this urge is applied to neonates of other species, for example, when people spend significant time and money to see, pet, or play with adorable young animals at a zoo or shelter. It can be difficult not to approach your own adorable and distressed child, even when it defies your own parenting philosophy. For example, even if your child cries less when you do not respond to a fall or injury, you may still feel compelled to rush toward them when they fall, in order to pick them up and soothe them in your warm and loving arms— even if it makes their distress last longer. The great conflict within ourselves and across parents of differing philosophies about whether children should always be retrieved or sometimes left to cry alone attests to the strength of the urge to approach a helpless child in distress and need.

Even though we possess a circuit in our brain that urges us to respond to infants or distress in general, there are differences in how each individual responds in each situation. For example, one reader of this chapter opined that the model could not be correct because that individual did not experience an urge to grab up or soothe a distressed child. Meanwhile, my teenager, who has yet to reproduce, spends hours each day looking at cute pictures of neonatal animals on the internet. Such varying reports show how the mechanism rests upon many interacting genes, which are affected by one's own unique genetic makeup and environment, producing a wide range of responses. For example, using the central limit theorem, Sir Francis Galton demonstrated in the 1800s that many genes plus the environment must encode human height.[23] Since people's height is normally distributed, with most people having middling heights but fewer and fewer people being extremely short or extremely tall, height could not be encoded by just one or two genes. If there were only one or two genes for height, you could predict any given child's height directly from that of their parents—which you cannot do, at least not precisely. As such, Galton demonstrated height must be determined by multiple interacting genes, impacted by additional variables such as diet and random nongenetic nuisance variables, captured by W.[24]

$$H = X_1 + X_2 + \cdots + X_n + W$$

Moreover, the fact that people gradually grew taller over decades and centuries, adding inches as people became more industrialized and were better fed, show that the environment affects the expression of genes for height (figure 5.2). (Remember, genes are sensitive to context by design.) Thus, just as people are biased to decide if a neutral facial expression is positive

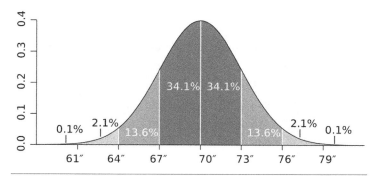

FIGURE 5.2 A histogram depicting the percentage of US men at each height, which is normally distributed, with most collecting in the middle values around 70 inches.

Drawing byCmglee, CC 2.5.

and smiling or negative and frowning, most people rush to aid a young child in great distress and need whereas others will insist that even children in clear need must figure out how to work it out on their own and still others will leap in to help with little provocation. These examples of variation—Victorian statistics on height and our varying sensitivity to children's need—demonstrate that even neural circuits designed to promote an urge to respond do not produce the same behavior in all people, or in one person over time. Even a neural circuit that we share with rats can produce complex responses that vary in sensible ways by individual, environment, and situation.

The Neural Circuit of Altruistic Responding in Humans

Whereas rodents are described as responding particularly to the smell and ultrasonic cries of pups,[25] humans are more likely to

see and hear the cries of a victim that are much lower in frequency, suited to our own auditory system. Some suggest that
the spectral frequency range of our auditory system evolved per
se to hear the cries of infants, given the importance to survival
(or course the causality could be reversed). Our auditory system
is highly attuned to sounds in the exact frequency of a baby's cry,
from 3 to 4 kHz. Anthropologist and opera singer William Beeman noted that this frequency also corresponds to the most
emotion-inducing part of a singer's vocal range, referred to as
the "singer formant region."[26]

After perceiving distress, multiple brain areas collaborate to
mark the event as important and to facilitate a response. A fast
perceptual route encodes less detail about what we perceive but
can activate the amygdala directly from the thalamus, without
bothering to first process the details such as exactly who needs
help and what their problem is. This fast brain activation from
the amygdala can then quickly prepare a response through direct
projections to brainstem autonomic regions that subsequently
increase heart rate and prepare our muscles to respond. This amygdala activation, having marked the stimulus as important, also
reinforces the slower neural route that continues all the while to
determine the exact nature of the situation, through routes pro-
gressing from the back to the front of the brain, along both the
top and the bottom of the cortex. This more gradual processing
is needed to determine aspects of the person, place, or thing that
the faster, more efficient process may have missed, such as the
identity of the victim or his or her precise location in space. This
slower cortical processing also allows us to place the victim's
distress into context with other events we have undergone, such
as linking it to memories of the individual or situation so that
we can tailor our response to characteristics that we have learned
from past experience.

After perceiving and identifying the victim's distress and need, the same brain areas that support offspring care in rodents also support the human altruistic response (figure 5.3). The MPOA was specific to pup retrieval in rodents and, thus, may not participate in human altruism unless the victim actually requires a physical retrieval—a heroic rescue, say. It will take time to confirm this potential role for the MPOA because it is small and, therefore, difficult to pinpoint using current human functional neuroimaging technology (fMRI). Because of this technical issue, we know little about the role of the MPOA in human behavior in general, and even less about it in the context of an altruistic response. In contrast, we already have evidence that the other brain areas in the circuit, such as the amygdala, hypothalamus,

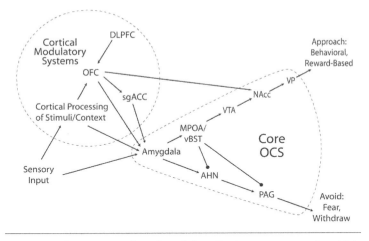

FIGURE 5.3 An augmented version of the offspring care system, in which I have added important known connections with frontal and cortical regions that participate in human altruistic decisions, referred to as the extended caregiving system.

Stephanie D. Preston, "The Origins of Altruism in Offspring Care," *Psychological Bulletin* 139, no. 6 (2013): 1305–41, https://doi.org/10.1037/a0031755, published by APA and reprinted with permission, License Number 5085370791674 from 6/10/2021.

and NAcc, participate in situations that involve infants, caregiving, and altruism.

Decisions Informed by Affect in the Frontal Lobe

During human decision making, information about the emotions, victim, situation, and possible outcomes converge in a part of the frontal cortex that sits just behind the eyes, in the front and bottom of the prefrontal cortex. This "orbitofrontal cortex" (OFC) integrates cues of the person and situation along with our emotional response to them in order to produce an advantageous response, all things considered.[27] The amygdala generates an initial affective response to the situation, while the hippocampus and other cortical brain areas encode associations that we learned about the person or event, in collaboration with the OFC and NAcc, so that we can place the person and situation into context based on our past experiences. The OFC also returns signals back to the amygdala (especially the basolateral nucleus, BLA) to influence our response to uncertainty.[28]

People do often sit back and consider their options when the situation is less urgent, which involves these interconnections between the more ancient regions that process emotion, motor responses, and arousal with the newer frontal lobe areas like the OFC and dorsolateral prefrontal cortex (DLPFC). In these less urgent situations, where people have time to compare possible outcomes in order to make an informed choice, these older and newer brain areas work together to bring possibilities into mind and to hold them there while we figure out which possible outcome feels the best so that we can select our response.[29] For example, when people try to determine which card deck to choose from during a gambling game, the DLPFC is necessary to develop a conscious, explicit understanding of the decks' relative good and bad outcomes.[30] However, just leaning toward a better over a worse deck does not require working memory and conscious awareness of the possibilities.

As an example of this implicit leaning process, Dan Tranel and Antonio Damasio studied an amnesic patient, "Boswell," at the University of Iowa Hospital. Boswell had suffered extensive damage to his medial temporal lobe. He could not remember faces, including those of the providers who had treated him for many years. Even his skin conductance arousal response did not look different when he looked at familiar versus unfamiliar faces. Despite his extensive amnesia, though, Boswell *could* distinguish between care providers who had treated him well versus poorly in the recent past. His skin conductance arousal response increased when he was faced with providers who had recently treated him well, and he chose those preferred providers from a lineup to ask for treat, despite having no explicit memory of meeting or knowing them or how they had treated him in the past. Boswell's brain retained an implicit, emotional memory of his interaction with those providers, which biased his choice. This is remarkable especially because Boswell also sustained bilateral damage to the hippocampus and amygdala—ancient brain areas that are normally required to create emotional memories. Perhaps then Boswell's intact prefrontal cortex and striatum could help him associate people with rewarding foods per se, especially because he was less successful at selecting who he liked or would help him.[31]

You can apply these decision processes to your own daily life. For example, if your neighbor came by to ask for a cup of sugar or to borrow your lawnmower, your NAcc, amygdala, and hippocampus would work together with your OFC to bring to mind any experiences with that neighbor and his past deeds. Had he previously called the police or brought an expensive bottle of wine to your open house party? The OFC, in concert with the DLPFC, would try to hold such memories in the forefront of your mind, to bias your decision as to whether to help the neighbor or not. Given this information, you would approach your neighbor in the yard if he had been friendly in the past and would surely loan him some sugar or your lawnmower—you might even run to his rescue when he struggled to carry packages to the door or shovel ice from the sidewalk. Your conscious deliberation about what to do in each situation would take even longer and be even more tedious if he asked you to provide a more costly or sustained form of help, such as taking care of his cat for a month while he traveled for fun. In this case, you might sit down and really think about what he has done for you in the past, how much you like him, how much this aid could contribute to your relationship, and how guilty you would feel if you declined.[32] This additional and sustained cognitive processing can also help people to inhibit an intuitive response that conflicts with their own long-term goals, and it allows people to respond even when they do not feel the urge if helping would benefit them. For example, even when your neighbor cries out from the weight of the couch he is struggling to push through his front door, you might not move a muscle to help if you knew he was overacting, had strong sons or tons of money to hire movers, or last month sat idly by in his lawn chair while you attempted to haul a refrigerator away.

Researchers and laypeople alike always think about these highly deliberated cases—when they thought long and hard about helping—even if such deliberations are not very frequent and very low-level instinctual processes can still play an important role.[33] For example, when you watch a sad television commercial asking for donations to feed impoverished children in Africa, the ancient offspring-care system might be activated when you perceive a young child who is clearly suffering and needs food to survive another day—despite having plenty of time in this case to carry out any decision. After you watch a moving television plea for help, you must still get yourself up off the couch, find your wallet, figure out the website or phone number where you are supposed to make the donation, and decide how much money would make you feel generous without affecting your ability to meet your monthly car payment.

Thus, even when people spend significant time deciding whether to help a neighbor and recall a specific event or imagine a possible outcome, the information and the way you think about it represent just the tip of an iceberg of a highly emotional and learned process that supports most decisions, much of which we share with other animals. As a simple "gut check," think of a time when the facts on your pros-versus-cons list clearly aligned with your first choice, but you chose the second one instead. For example, there are a million reasons (or so) that I should go kayaking at the end of each day and perhaps only one con, yet I rarely go. For example, I already have a stated goal to exercise more, to use the kayak that I paid good money for more often, and to enjoy nature as a way to relax and savor life. But, somehow, at the end of every day, when I consider whether or not to kayak, I end up sitting back with a beer or a snack because, all things considered, kayaking just sounds so tiring.[34] Thus, even explicit, calculated, and rational decisions are undergirded by implicit emotional associations, predicted consequences, and the value that we place on each attribute—information that is hard to render on a simple pro/con list. Similarly, even the human altruistic response is strongly influenced by emotional and subcortical processes that we share with other species, which are not always conscious, and that are piqued by immediate rewards like feeling warm and safe after a quick rescue or feeling relaxed after a cold beer.

Evidence from Human Neuroeconomics

If there is indeed a homology between offspring care and human altruistic responding in the brain, then the brain areas that support offspring care in rodents should also be activated during human altruism. To date, the evidence is sufficient, albeit imperfect and indirect. After all, it is hard and not particularly beneficial to modern medicine to replicate heroic rescues while lying motionless in a loud fMRI scanner. Despite these limitations, there are many consistent demonstrations that the brain areas supporting offspring care also participate in human decisions to help, even in situations that are not that similar to pup retrieval, such as when an adult gives money to another adult stranger who does not have an urgent need (e.g., the OFC, NAcc, insula, amygdala).

Most altruism experiments in psychology involve behavioral economic games, in which a subject (usually a student) is given an amount of money by the experimenter and then decides whether and how much of that money to donate to or trust with a stranger (usually another student) or keep for themselves.[35] In the "ultimatum game" the subject in the experiment can offer any amount of their new money to a stranger, who can then accept or reject the gift. Objectively, partners should never reject free money, but they often do, which economists interpret as evidence that we possess a general bias to be cooperative and punish those who are not.[36] When participants think that the offer that the first person gave them was unfair, brain activity increases in the insula and ACC, which are thought to track negative emotional feelings that are associated with being treated unfairly. During this game, brain activity also increases in the DLPFC, which is thought to inhibit the recipient's desire to reject an unfair offer so that she may benefit from the windfall, even if it seemed unfair.[37] For example, when researchers block brain activity in the lateral frontal cortex (including right DLPFC) using transcranial magnetic stimulation (TMS), the recipients are less likely to punish an unfair offer, even though they recognize them as unfair.[38] Conversely, when people sustain brain damage to a more ventral and medial portion of the prefrontal cortex (VMPFC, which is like OFC but larger and less defined) they reject even *more* unfair offers.[39] Thus, perhaps the lateral portion of the prefrontal cortex inhibits reward-seeking processes from the striatum, whereas the ventromedial region in an intact brain ameliorates a shorter-term desire to punish that is fueled by disgust-related processes in the insula and ACC, allowing us to obtain the longer-term rewards of benevolence.

In the "trust game," subjects are given money and can trust some of it to a stranger as an investment. The stranger's new gift is then multiplied by the experimenter, and the recipient can return as much of this new, larger amount back to the original subject as he or she wishes, including none. Subjects who trust their partner with their initial allocation have more brain activity in the prefrontal cortex when they are waiting to learn their partner's decision, compared to people who didn't trust the partner or believed the partner was a computer.[40] When both partners in the game are scanned at the same time, brain activity increases in the cingulate, septum, VTA, and hypothalamus—usually when trust was given or received. Supporting the altruistic response model, the septum and anterior hypothalamus, which control the expression of oxytocin and vasopressin, are engaged during trusting, positive partnerships in the game.[41] However, the cingulate is more necessary to predict the partner's response before deciding, which is unnecessary when partners always trust each other. The VTA is also more active when one was defected *upon*, and in pairs with low trust and reciprocity. Taken together, trusting relationships appears to bias behavior toward cooperation, with less cognitive effort and more social boding, whereas unstable relationships require more consideration about what the other might do and the level of reward on any given trial. In a positron emissions tomography (PET) brain imaging study with this game, the proposer could punish a partner who defected and did not return any of the multiplied gift. In this situation, brain activity increases in the dorsal striatum (caudate) and thalamus when participants punish the defecting partner (versus when not punishing or when punishment was symbolic); activity in the caudate even correlates with the amount that the jilted partner would pay to punish them.[42] When using their own money to punish a defecting partner, brain activity increases in the medial prefrontal cortex (OFC, VMPFC), perhaps suggesting that these recipients integrated their competing goals toward a satisfying choice. Thus, altruistic punishment appeared to reward the punisher, even though it cost them money.

In a "prisoner's dilemma game," both partners receive the most money if they each separately choose to cooperate, but neither of them gets any money if they both defect, and if one defects and the other cooperates the defector gets everything. When two women choose to cooperate on this task, brain activity increases in the OFC and NAcc, which the researchers interpret as a sign that working together is rewarding and reinforcing.[43] When one's trust is subsequently rewarded with cooperation from the trusted partner, brain activity increases in the VMPFC and ventral striatum, more than when the trustor experiences a defection.[44] In a similar game, participants in the brain scanner played against a fake partner who acted fairly or unfairly before receiving (again fake) electrical shocks to their hands.[45] As with Boswell the amnesic patient, fair partners are perceived positively (more agreeable, likeable, and attractive), and participants feel more "empathic pain" when they have to watch their fake, fair partner be shocked (i.e., activation increased in the insula and ACC that represent felt pain) compared to when the unfair partner is shocked. This effect is strongest in women and in people with empathic personalities. In women, the empathic pain responses in the insula and ACC are lower for unfair than fair players, but the empathic pain response is lacking when men observe unfair partners getting shocked. Moreover, only in men does activity increase in the NAcc and OFC when an unfair player was shocked, which the researchers interpret as a sign that the men enjoyed their pain (i.e., "schadenfreude"), because it correlates with their desire for revenge.

Taken together, people develop emotional associations with others through accumulated experiences, which then feedforward to influence how they respond to social partners in an hour of need. We like and empathize with people who cooperate with us, and we dislike and feel less for those who let us down or take advantage of us. As such, whereas we might feel the urge to rush toward a neighbor who fell on the icy sidewalk after he brought us a nice gift (and may even shovel it for him henceforth), we may snicker in retribution from the warmth of our living room while watching him fall if he previously called the cops to break up our party.

Evidence from Measurements of Oxytocin

These studies were presented to explain how the brain areas associated with offspring care and that emotionally inform decisions also support human decisions to give. Substantial research also supports the homologous role for oxytocin in human altruism, often using the same behavioral economic games as before. For example, oxytocin increases in the blood when participants trust their partner and return more of the money back to the giver, especially when the exchange seems intentional (above and beyond the sheer amount transferred).[46] When oxytocin is administered through a participant's nose before a trust game, the first person gives more money to their partner, even more so if they receive a massage beforehand,[47] which could have upregulated oxytocin through the deep, personal touch. The length of the promoter gene that encodes for the expression of oxytocin is also longer in participants who give more to a stranger in the dictator game.[48] In a functional neuroimaging game, participants who receive oxytocin do *not* trust their partner less after experiencing a defection, which is associated with reduced brain activity in the amygdala, midbrain, and dorsal striatum—areas that support offspring care and reward-based decision generally.[49] Oxytocin administration also decreases the amygdala response to another's pain, without any direct correlation among oxytocin, empathic pain, and monetary donations.[50]

Some of these experiments that involve oxytocin have not been replicated and statistical analyses that combine effects across many similar studies report that the impact of oxytocin is small and may not differ from zero or may affect performance only on some measures (e.g., facial emotion recognition or higher trust for in-groups over out-groups).[51] As a guiding principle, given the evolutionary origin of oxytocin and the contexts in which it acts across species, one should only expect oxytocin to support behavior when there is a social bond and not assume it will be involved in unnatural laboratory situations that involve other types of rewards. In contrast, the NAcc appears easier to engage across situations, as it is involved in any choice that includes a motivating reward, and not just with social bonds. We should test explicitly whether oxytocin is more engaged by contexts that involve caring and bonding over abstract financial transactions with strangers (e.g., an interpersonal exchange with a close other versus giving "house money" to a stranger).[52]

Evidence from Human Charitable Giving

A few studies have examined neural activity during human decisions to donate to charities. This context is more similar to offspring care compared to donating house money to other affluent college students, even if it is still not heroism or a literal rescue. At least in charitable donations, the recipient is described as having clear need that may pull more at the heartstrings.

One study compared charitable giving depending upon whether the subject also received money with the gift (less altruistic) or just gave it away (more altruistic), and whether their donation was mandatory, as with a tax (less altruistic), or voluntary, as with a gift (more altruistic).[53] When the charity and the subject *both* receive money, brain activity increases in the dopaminergic ventral striatum, and voluntary givers with striatal activation donate twice as much. When subjects give voluntarily, brain activity increases in the caudate and right NAcc and participants are more satisfied with their gift, supporting the idea that people are rewarded by and feel the "warm glow" of giving when the gift is genuine.

A similar study asked participants to allocate either their own or the experimenter's money to a variety of real charities aligned with opposing political ideologies.[54] In this context, both receiving money and donating it to a charity activates the dopaminergic mesolimbocortical system, including the VTA and dorsal and ventral striatum. This activity also correlates with how proud and grateful the participants feel. Ventral striatum activity even increases as the donation becomes more costly to them. More selfless or costly donations generally produce more anterior activation (e.g., frontopolar and medial frontal cortex), correlated with the amount of money participants donate to charities in real life. When subjects donate over received money, activation increases in the subgenual anterior cingulate cortex (sgACC), which is also involved in feeling guilt from harming someone in another study (figure 5.4).

Research from across domains suggests that this particular region—the sgACC—is required for an effective altruistic response because it helps regulate emotion and the parasympathetic nervous system during sad or distressing situations. The sgACC is extensively interconnected with other regions in the offspring care system and with reward-based decision areas such as the OFC, lateral hypothalamus, amygdala, NAcc, subiculum, VTA, raphe locus coeruleus, PAG, and nucleus tractus solitarius (NTS). Displays of need typically involve sadness or distress, which should activate the parasympathetic nervous system and the sgACC. For example, the sgACC is activated when mothers listen to the cries of their infants, making it a good candidate for supporting altruistic responses to distress in people who are close to us.[55]

In our laboratory, we tried to more directly test the altruistic response model by presenting participants with charitable causes that they could donate to, with money they earned during the study through a finger-tapping task.[56] Unbeknownst to participants, the charity descriptions contrasted victims who were neonates or adults who needed immediate aid or just preparation for a later possible situation, and the aid could take a more nurturant or heroic form. As predicted, participants prefer to donate to young victims, when the aid is required immediately and requires heroism. But in an unexpected three-way interaction, the highest

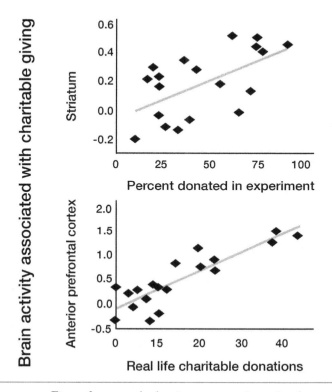

FIGURE 5.4 Figure from a study showing greater striatum involvement when participants donate more money to charities and anterior prefrontal cortex engagement during the task for people who donate more in real life. Both regions are central to the extended caregiving system.

Redrawn by Stephanie D. Preston from information in J. Moll, F. Krueger, R. Zahn, M. Pardini, R. de Oliveira-Souza, and J. Grafman, "Human Fronto-Mesolimbic Networks Guide Decisions About Charitable Donation," *Proceedings of the National Academy of Sciences USA* 103, no. 42 (October 17, 2006): 15623–28. Copyright 2006 by National Academy of Sciences, U.S.A.

donations go to young victims who immediately needed *nurturant* aid. This peak in the donation of their own money is associated with brain activity in multiple areas that are required to plan and enact motor responses—regions that are also involved during a motor-reaching task in the same experiment that did not involve people or money. As such, we demonstrated that victims and situations that resemble an offspring care situation also promote altruistic responses to strangers, as predicted by the altruistic response model.

SUMMARY

The preceding studies largely support the altruistic response model, inasmuch as the same brain areas that support offspring care are also engaged when people act altruistically, even in experiments that are not really anything like feeling an urge to retrieve a helpless neonate. The closer the context came to that of an infant in need, the more the response employed ancient, subcortical brain areas that are known to support offspring care in rodents (e.g., hypothalamus, sgACC, brainstem). Most of the experiments were able to demonstrate the role of reward-based decision-making areas during decisions to help another person (e.g., the ventral striatum, VTA, and NAcc)—particularly when people relished their response, whether it was a gift for themselves or a punishment to someone else. The OFC, in contrast, was more involved in situations where the decision involved conflicting responses that the participant had to integrate into a choice. This is consistent with the idea that more natural decisions to give, where the rewards align with one another, can be handled by ancient subcortical processes whereas more deliberated decisions require inputs from these ancient caregiving regions (e.g., NAcc, ACC, insula) into portions of the frontal lobe that help people make informed decisions in more complex and slowly unfolding situations. Given the diversity of brain areas engaged across studies, there is clearly no "altruism area" in the brain. Instead, the relative amount that any given brain area is active depends upon the study, the task, and the individual.

Virtually none of the studies have examined types of aid that would have been available long before industrialization and the invention of abstractions like money. Of course, there are real-world parallels to explicit decisions to give money—choices that are much easier to study with the neuroimaging methods that we currently have available. Charitable donations are a little like our ancestral form of giving because they at least involve a victim who needs our help, and there is some measurable motivation by us as observers to respond. However, even these studies require conscious deliberation more than they allow for an active urge to help. When someone rushes toward a victim in need, the decision is more straightforward, the costs are not explicit, and the giver does not need to sacrifice any of their own rewards for the recipient.

According to the altruistic response model, people feel the urge to rescue victims only when they feel competent and predict success, conditions under which an observer could even save a life without incurring too much risk to themselves (other than the bother of fame perhaps). When we approach a lost toddler at the mall, help a neighborhood child get back on his bike, or reach out to support the bus passenger about to fall, we are not giving up money or making any substantive choice whatsoever. Such acts are only decisions in the nominal sense that all motor acts are (because one action was selected among multiple conceivable options); however, the alternatives to helping in daily cases like this do not need to be salient in the mind of the helper. The more the helper's response resembles the retrieval of a helpless infant, the more it rapidly dominates processing without presenting itself as one of multiple options to consider.

The fact that there are other alternatives during a prototypical act of altruistic responding is something that is characteristic of the situation, not the mind of the giver—a distinction that is crucial to the altruistic response model and that is greatly underappreciated by existing models of decision making, morality, and altruism.

It is even possible that the controlled, cognitive processes that are required for most economic and neuroimaging studies of human altruism *inhibit* people's natural motivation to help, because it disengages them from their natural drive state. For example, when researchers compared altruistic responses in children and chimpanzees, the subjects did *not* help more when they were given a reward for helping; in fact, twenty-month-old children helped *less* when rewarded for helping, presumably because the money offset the warm glow that only follows an intentional and genuine gift.[57]

Neuroscience needs to create ways to examine a direct, immediate response to assist someone in clear distress and need, with real aid. Perhaps variations of the classic bystander paradigm or nonhuman studies where one animal terminates the distress of another can be used. Such tasks would be expected to activate more posterior and medial regions in the brain (e.g., amygdala, NAcc, and sgACC) over the frontal lobe areas that are required for more explicit choices (e.g., frontal pole, DLPFC, ventromedial PFC). Additionally, research should determine the degree that explicit deliberation can block or inhibit an urge to respond, given that abstract, monetary tasks that require a trade-off currently dominate our knowledge even if they do not generalize to the types of real-world aid that we evolved to give.

There have not been direct tests of the altruistic response model in psychology and neuroscience, but the aggregated research on human altruism in neuroscience has produced convergent evidence that brain areas that involve dopamine and reward-based decision making (e.g., OFC, amygdala, hypothalamus, NAcc, sgACC), which are also modulated by oxytocin, participate, as they do in offspring care. Future work can more directly test whether the specific features of a helpless, distressed neonate in need foster an urge to help and engage the offspring care system, especially when compared to more rational, deliberated cost-benefit decisions.

6

CHARACTERISTICS OF THE
VICTIM THAT FACILITATE
A RESPONSE

Related or not, infants can be powerful sensory traps,"
writes the eminent anthropologist Sarah Blaffer
Hrdy.[1] Because of this pull that we evolved toward
infants, people and situations that mimic our helpless offspring in some way can also drive us to attend to and approach
them. There are four main features that influence how we perceive a victim, owing to our ancestry as a caregiving species,
that predict whether we will feel the urge to help someone
in need, even an adult or complete stranger. These attributes seem straightforward enough perhaps, but I will define
and clarify each one in turn so that we may understand how
each operates in isolation and also interacts with the others
(figure 6.1).

- Vulnerability
- Needing immediate aid
- Resembling a newborn or child (neoteny)
- Displaying distress

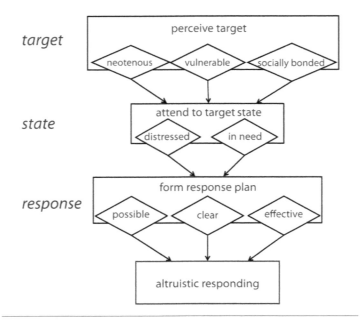

FIGURE 6.1 Flowchart depicting factors that can release the altruistic urge and that predict a response. Even if all these factors combine during offspring care, they can trade off for one another in a continuous, additive manner during human altruistic responding.

Stephanie D. Preston, "The Origins of Altruism in Offspring Care," *Psychological Bulletin* 139, no. 6 (2013): 1305–41, https://doi.org/10.1037/a0031755, published by APA and reprinted with permission, License Number 5085370791674 from 6/10/2021.

VULNERABILITY

The first key entailment of a caregiving-based model of altruism is that the motivation is most strongly directed toward victims who are *vulnerable* in much the same way that offspring are. Vulnerability enhances our sense that the victim really is in danger by causing us to sense that the victim is less able to handle the problem alone and need our help. Of course, babies are naturally vulnerable; their prolonged immaturity prevents them from taking care of themselves, putting them at serious risk to threats like starvation or predation—the reason we evolved this instinct in the first place. Even though adults are not usually considered vulnerable, they can be rendered so during particular conditions, stages of life, or emergencies. The altruistic response is potentiated in such cases, where an adult victim is rendered vulnerable by their situation, such as during a heroic rescue. For example, in the most famous modern case of heroic altruism, Wesley Autrey was observed in New York City jumping onto a subway track to rescue a young man who had fallen onto the tracks just as the train approached.[2] The onlookers had seen the victim having a seizure just before he fell into the tracks, presumably the reason for his fall. Likely those observers, including Autrey, understood that he had suffered a neurological event and was incapable of extricating himself from the train. In that moment, the young man was rendered vulnerable in a way that was not characteristic of his age or degree of maturity, but instead reflected an acute, urgent problem that was beyond his control. In combination, these factors facilitated Wesley's immediate and even risky response, to save the imperiled young man.

In a mundane example, engineers assist vulnerable people through the design of our city streets. As a parent, I have spent a lot of time waiting at intersections and shouting, "Walking Man!" to my kids so that they know it is safe to cross the street, referring to the white human-shaped form in the signal. Traffic engineers refer to children, the elderly, and people with disabilities as "vulnerable users" of our city streets, because they are at increased risk of being hit by a car when crossing the road. An able-bodied adult might find it relatively easy to detect a gap in traffic and dart across the street, or they may have no problem reaching the opposite curb in the time that the brief Walking Man affords. In contrast, a vulnerable user might have trouble deciding when to cross a busy street or may be too slow to reach the opposite curb—with or without a crosswalk. Engineers add time to crosswalk signals to accommodate diverse users and add signage and crosswalks where vulnerable users occur (e.g., noting where a blind or deaf person lives or adding user-activated signals at midblock crosswalks to stop cars where children cross to school). Typically, vulnerable users suffer from chronic issues (being a child is also somewhat chronic); however, sometimes acute issues render someone temporarily as vulnerable as a baby, such as when suffering from acute illness, injury, neurological event, or unconsciousness. Like the young man who had a seizure in New York and needed immediate assistance, this acute vulnerability was probably not typical for him, but it was recognized by onlookers, which precipitated an urge to respond (assuming other features of a victim, which follow, do not contradict the urge).

One unfortunate corollary of this by-design link between infants, vulnerability, and aid is that many people in great need do not receive help because they do not appear vulnerable to us, and, conversely, we can direct aid to those who seem vulnerable but do not actually need our help. These complications are addressed in what follows.

Complications to Our Perception of Vulnerability

Because vulnerability is tied to our urge to respond, people can sometimes seem apathetic or callous in ways that are unfortunate but predictable from the altruistic response model. For example, people are less prone to rush toward a victim who appears to have been responsible for his or her plight—if, say, the young man had fallen onto the tracks because he was intoxicated instead of having suffered a seizure. This perception of responsibility dampens the response. The urge would still exist if the victim were truly young and helpless in the face of temporary, urgent danger, such as if it had been a thirteen-year-old boy who had fallen onto the tracks after passing out from intoxication. People differ in the degree that they penalize victims who seemed to have caused their plight, a rationale that people use to justify inaction that is statistically linked to political ideology—for instance, with political liberals being less punitive than conservatives.[3] It is often hard to appreciate how someone arrived at a place where they put themselves into acute danger or need, which may have emanated from a condition like addiction or poverty; it is also hard to understand how difficult it is to exit such a situation without help. For example, if the intoxicated boy who fell onto the tracks were raised by a single parent who was an addict, had a genetic predisposition to alcoholism, and was grieving a lost friend, he would be suffering under the weight of tremendous issues that would affect anyone, which were not necessarily his fault; however, the average passerby could not know this. Thus, our apathy for a victim's plight may not accurately represent his situation, and our urge to respond is highly tied to what we can observe directly or assume with some additional perspective taking.

People are also more likely to help vulnerable victims because they seem more likely to accept and appreciate help. A truly helpless victim cannot fix the problem without an intervention, but a stranger in a public place who is not clearly vulnerable could be offended by the aid that undermines his independence or mastery.[4] The natural association between vulnerability and youth also means that people who receive help, can assume that they were viewed as vulnerable or needy, which makes them feel

patronized by the helper or treated like a younger, more subordinate, or less powerful individual.

It is surprisingly difficult to determine whether people really
need and want your help. For example, an intoxicated person may
need help but not want it; that person may even aggress belliger-
ently against you for offering help. Imagine if the chronically
intoxicated boy of our prior example were your nephew. He might
ask you for twenty dollars to buy something for the day. You might
refuse this aid even if it was not a large sum for you and even if
you would pay much more for inpatient rehabilitation because
you love him and want him to thrive. However, your nephew may
not want to go to rehab; he just wants cash enough to survive
another day. These are complicated situations. In such cases, there
are many attributes of offspring need that may facilitate an urge
to respond (e.g., vulnerability, need, youth, love), but if an addict
does not want your help, or does not want the type that you prefer
to give, he can become a drain on your energy and finances. These
cases exist in a gray area where some prefer to let the nephew
figure it out on his own or reach "rock bottom," whereas others
will pay whatever the young man wants because it is too painful
to watch him suffer. People shift toward reticence if the addict is
no longer young, shows no signs of changing, or uses the money to
further endanger his health. Thus, even when vulnerability and
need are clear, aid may not follow if the victim does not want the
help and other cues can conflict with a response.

Someone with a physical disability may struggle while opening a door even if he or she can manage the task and prefer to be
left to it. In the twentieth century, men in the United States were
taught to open doors for women as the gallant thing to do. Some
women feel patronized by this aid, because it implies that women
are weak, incompetent, or inferior to men, even if the aid was
well intentioned. Recently, I passed a woman in a wheelchair

entering a conference and considered opening her door, but the woman seemed capable and accustomed to this process, even if it was harder and took longer for her. Help can rob people of the pride of independence that they feel from doing things for themselves and make them feel intruded upon or "less than." Because the woman I passed did not seem truly helpless and I feared that my aid might irk her, I decided not to intervene. I will never know whether the aid would have been appreciated or not, but this example is presented to show that we have an urge to help people who appear to us as if they are struggling and vulnerable. But there is a large gray area in the continuum between obvious need and clear competency. We are very sensitive to these cues, but uncertainty promotes inaction when they conflict. Regardless, when a truly helpless individual needs urgent help—like the young man who fell onto the subway tracks after a seizure—we can feel the urge to help.

Chronic need can also dampen the urge to help in situations such as those of caregivers who habituate to their charge's need or who feel burned out, such as the spouses of elderly or infirmed relatives or care providers at a facility. For example, when someone suffers from a disease like Alzheimer's, Parkinson's, multiple sclerosis, arterial lateral sclerosis, cerebral palsy, or paraplegia, their vulnerability and need is clear, but they can still have needs that go unattended if the caregiver is used to the need and does not view it as urgent, like changing clothing or bedding, being taken to the bathroom, or being bathed. These acts are necessary for a healthy life but are not technically required *immediately*, as being pulled from an oncoming train is. As a result, these less urgent needs do not precipitate the same urge to respond.

My father had Parkinson's disease and needed help with everything in the last few years of his life. Certain mundane tasks were required many times a day because he had problems using and keeping track of things such as finding his glasses or the television remote or reconnecting his tablet to the internet. These problems were significant for him because they involved the few remaining activities that he could do, like watching television or reading the news on his tablet. Even so, the constancy and mundaneness of the tasks could fuel our impatience or slow my response at times. My father could also be quite impatient himself when making these requests for help, which could annoy us despite understanding his disability, lack of control, and need for such frequent infusions of aid. Our urge to respond was dampened by fatigue and a lack of urgency, despite his vulnerability and the fact that we loved him very much.

That chronic need reduces the urge to help is particularly regrettable in cases such as placing loved ones in nursing homes, where elder abuse is common and hard to eradicate without constant oversight.[5] Even people who are naturally caring can habituate to and resent the routine (but real) needs of their charges. Care providers can become drained by constant requests for help, particularly for tasks that are not urgent or glamorous—certainly not as glamorous as leaping before a subway train in front of a crowd of onlookers. Some chronic aid is downright unpleasant. There are also few awards or ceremonies to honor those who selflessly change bedpans, wash soiled sheets, or clean bathrooms. Even though these chronic issues place limits on the altruistic urge, they can be addressed if we understand and apply the theory. For example, in a long-term care facility we must regularly point out the perspective of the patient, who feels helpless, degraded, and abandoned—and who will be you someday. We should yoke aid tightly to a schedule, like providing medicines, meals, baths, and social engagement at specific times to bypass a reliance on a motivation to help altogether. We can provide private and public recognition for such work, no matter how mundane. Certainly, higher pay is warranted for jobs that are necessary, roundly unpleasant, and that beloved family members could no longer do at home by themselves.

Caregivers can sometimes even negotiate with patients on decisions such as what the caregiver is willing or able to do and how long it should take. For example, I now fully understand why my father was so frustrated by these small problems that were important to his quality of life. We probably could have saved ourselves some trouble if we had just negotiated how long we would take to respond to such requests (say, ten minutes), rather than just retorting each time, "Okay! I'll be there in a minute!" My dad could have engaged in such a discussion about his care, but it was hard for him to inhibit himself when he really wanted something (and dopamine medication made him even more impulsive). Poor inhibition is actually an additional form of vulnerability that patients have little control over, which warrants our empathy and patience. But in the throes of caregiving, grief, and confusion, distanced reasoning is hard. Through the altruistic response model, we can understand how these types of cases conflict with our evolved urge to help and, thus, can focus more on our love for the person and plan a response that allows us to provide the compassionate care we want to give, even without an urge.

IMMEDIATE NEED

The prior examples about vulnerable victims often included urgent need, which leads us to the second characteristic: immediacy. Neonates are inherently vulnerable, but their need is sometimes urgent and sometimes not. The two attributes act independently in the case of infants, and either can separately or together fuel the mother's response. Vulnerable adults sometimes need immediate aid, such as the young man who fell onto the subway tracks before an oncoming train. In general, heroic rescues involve both vulnerability and immediate need, as was the case for the pups that needed to be retrieved. Thus, vulnerability and immediacy are not the same attribute, but they often co-occur, especially in canonical cases of the altruistic response such as a rescue.

We are the most motivated when the victim needs our help *right now*, not "if and when we can get around to it." We often observe someone's genuine need and even feel empathy and sympathy for them without managing to get around to providing that aid. We want to help. We plan to help. But when the need is not urgent, we assume that we can address it after we finish this other thing, at some vague future point that remains just out of reach, such as when we have more money in the bank or more time on our hands. Thus, sometimes victims really are vulnerable and want our help, and we want to help them back. But because the need is not urgent, a convenient time never seems to arrive, and we simply fail to act (though we might still congratulate ourselves for thinking of them). We might even justify our inaction by minimizing or rationalizing their need so that it seems less urgent or important—a sad irony, since people often feel happier after they do help someone. This gap between feeling for someone and actually responding is what Tony Buchanan and I call the "empathy-altruism gap."

These conflicts between our hopes and goals and our actual responses can reflect what economists call *delay discounting*. In delay discounting, people commonly sacrifice a larger monetary reward in the distant future for a smaller one that is available sooner—especially right now.[6] Shorter-term and immediate rewards loom much larger in our minds because we can easily imagine how we would spend that windfall today, whereas it is harder to imagine what we will be doing or even who we will be later, when the larger reward finally arrives.[7] Our natural bias toward more immediate rewards is usually bemoaned, likely because Western, industrialized people are always striving toward those longer-term rewards, such as obtaining a college degree, saving for retirement, and eating a salad today to be thinner next month. For example, young people might plan to save money for retirement but fail to put anything away each month because they can always think of a concrete item that they need right now. An academic might spend each day dutifully responding to emails from students and colleagues and never finish that book that she cares so much about, which would benefit her career more in the long run. A person might want to lose weight this year but will watch television with popcorn each night instead of exercising, even though he really does want to be more fit. The activities that we select in the moment are usually easier to imagine, easier to take part in, and feel good sooner. In contrast, goals like writing a book or getting in shape are abstract, distant, and take longer, making it hard for them to compete for our attention.[8]

The same is true of altruism. We might fail to help someone who does not seem to require aid *today* in favor of a more urgent-seeming task like replying to emails, doing the dishes, or buying a latte. Over many years, these small decisions add up. Later we may regret how little we helped those close to us or how little pride we can take in our altruism. Our preference for the less valued but urgent-seeming tasks—say, to be fit or altruistic—is all the more regrettable because people actually *enjoy* exercise and helping others. For example, people in one psychology experiment reported being happier to the degree that they spent more on others, gave more of a work bonus to others than themselves, and spent a $5 or $20 gift from the experimenters on someone else or a charity rather than themselves.[9] Thus, people were happier after spending money to improve someone else's life, even though they could have instead rewarded themselves with that latte or a nice lunch. This phenomenon is linked to the "warm glow" of giving, in which people feel good knowing that they helped someone else.[10] This warm glow is powerful because it is built into our brains through the system for caring for offspring, like the dams that seemed just as rewarded by having to retrieve strange pups that fell from the sky, despite the added effort.

Even if we bemoan this bias toward immediacy, it is largely adaptive. All caregiving mammals need to be highly attuned to the cues of distress and immediate need in order to ensure the safety, security, and survival of their offspring—right now. Moreover, acute need is often the most life-threatening. Thus, just as the emergency room uses a triage system to prioritize patients who might die over those with a broken bone, our brains prioritize victims who are facing more dire consequences. This is usually a good thing.

Our brains are designed to favor responses to immediate rewards and problems, whether we are talking about money or food or people in need. This bias is generally adaptive and understandable, but it also means that the "walking wounded"—people with chronic need that is debilitating and real—are often ignored. Our bias toward immediacy can also create problems with how we prioritize or allocate aid, such as motivating us strongly toward a single child needing immediate aid over an adult who is in much greater danger that is less obvious or less urgent, as Paul Bloom points out in his diatribe against empathy.[11] By understanding the way that our altruistic urge evolved to respond to immediacy, we can explain confounding cases where people fail to act—even in accordance with their own expectations or goals—and, if we are lucky, do something about it.

NEOTENY

Neoteny is a term from biology that refers to the fact that offspring possess features that promote care because they are associated with a longer period of early development, which requires our extensive care before independence (i.e., the longer "altricial" development as opposed to the shorter "precocial" development). Neotenous features in infants, known as *Kindchenschema* in German, are thought to be attractive so that we attend to infants and feel motivated to approach and care for them.[12] For example, Konrad Lorenz established that offspring across species possess a relatively larger and rounder head than adults, along with a shorter nose and limbs (figure 6.2). Demonstrating the cross-cultural relevance of this phenomenon, the Epio culture in Micronesia has a specific word (bico) that refers to creatures like babies and small mammals that are cute, sweet, need protection, and motivate us to approach and cuddle them.[13]

Many Westerners are familiar with the Japanese term *kawaii*, which refers to an emphasis on being cute, through features like the large round head of Hello Kitty or the baby-doll dresses worn by young women. The meaning of the term *kawaii* has morphed over time such that it now refers to a kind of attractiveness with a desire to stay with and care for the target of attraction, a meaning that surely parallels Konrad Lorenz's and Robert Hinde's descriptions of how infant cuteness compels our care.[14] The concepts of beauty and cuteness may be particularly intertwined in Japanese culture, for the word for beauty (*utsukushii*) originally had a meaning very similar to *kawaii*.[15]

FIGURE 6.2 A depiction of the way that infants, across species, possess "neotenous" features, such as larger and more rounded heads and eyes, which are thought to facilitate caregiving.

Drawing by Miguel Chavez, CC-BY-SA-4.0. Redrawing based on Konrad Lorenz, *Studies in Animal and Human Behaviour: II* (Cambridge, MA: Harvard University Press, 1971), 155.

The urge toward adorable neonates is surely strongest for one's own bonded offspring, who are concurrently neotenous, vulnerable, and bonded to us. These features evolved to act as sign stimuli, which can then sometimes release an altruistic response toward even unrelated neonates, strange adults, or consumer goods. For example, people are known to be more likely to help adults who possess neotenous features and they are more lenient when neotenous children and adults make mistakes; people also consider adult faces to be more attractive after being morphed to look more neotenous, but they also consider them more submissive, incompetent, weak, and in need—features that are naturally associated with neonates and that facilitate a response but that are not uniformly valued in adults.[16] In one study, the experimenters made it seem as if someone had mistakenly left their résumé in a public place, such as the food court at a shopping mall, tucked inside of a preaddressed, stamped envelope. Half of the résumés included photographs that were morphed to look more neotenous, and half less so. As predicted, strangers were more likely to mail in the lost résumés for applicants that had the infantlike features.[17]

In the most recognized photograph in the history of *National Geographic*, Steve McCurry depicted Sharbat Gula, a beautiful Afghan girl living in a refugee camp in Pakistan during the Soviet occupation of Afghanistan.[18] Sharbat stares hauntingly into the camera, with visible dirt on her face and a ripped robe, with her iconic, large, light-green eyes that seem to demand that we attend to the plight of refugees. This image is perhaps powerful because it activates several of our cues of need simultaneously, such as being a young woman with striking neotenous features who is also in clear need. The cover photo from 1985 has since come to symbolize the need of refugees everywhere, who deserve our sympathy and aid, even when they are from a different nation or religion.

Our predisposition toward neoteny can also be manipulated to elicit attention and help. For example, Bugs Bunny used to be depicted in a dress and makeup while posing as a damsel in distress who manipulated others through exaggerated, childlike behaviors, such as a diminutive pose, a sympathetic pout, and widened eyes that he emphasized with slow blinks. The 1990s *Cape Fear* reboot by Martin Scorsese similarly includes uncomfortable scenes where the villain Max Cady (played by Robert De Niro) tries to seduce his enemy's high-school daughter, Danielle (played by Juliette Lewis), who highlights the inappropriateness of the moment and her own sexuality through similarly exaggerated, childlike acts. In South Korea, women and men undergo expensive facial surgery to increase the wideness of their eyes and reduce their chin or jaw, which enhances attractiveness through a partially more childlike appearance.[19] Cues of neoteny such as disproportionately larger eyes and heads have been increasingly exploited in modern culture to make toys or animated characters more salient and attractive. It has been suggested that the facial expression of fear was shaped over evolution to attract aid and attention because the widened eyes exhibited by frightened people mimic the enlarged eyes of a helpless baby.[20]

Neoteny is distinct from vulnerability, distress, or immediate need because it is more fixed and does not usually appear in adults. Neoteny and vulnerability are inherently linked in actual newborns. As individuals age, their features become more mature, with thinner faces and longer noses. Because of this dissociation, our altruistic urge is the strongest toward actual neonates, who are simultaneously vulnerable, neotenous, and adorable and need our help. Neoteny can facilitate aid toward adults, but it should affect our response less than vulnerability, because vulnerability is more tightly linked to the actual need for help. We might be attracted to and motivated by an adult's large, rounded eyes and view them as more helpless, but this would not precipitate a true rescue urge if they were not also vulnerable and in urgent need. For example, if Dwayne "The Rock" Johnson fell onto the subway tracks after a seizure, I would still panic and feel an urge to help, even though he is usually quite attractive, competent, and strong (of course, I would need to find help to lift him out). Perhaps Johnson is also appealing because his large eyes, rounded and bald head, and socially naïve characters foster a neotenous percept. Thus, being young or cute is not *required* to elicit an altruistic response, but it does contribute to the urge.

There are specific times when we do not want to help someone even if they are young and helpless. For example, people in America generally avoid approaching unrelated infants, even distressed ones who need help, if someone more qualified or related is nearby. Parents fear and reproach strangers who approach or touch their children, being uncertain of their intent. We tell stories and make analogies to "mother bears" that attack people who pass between her and her cubs. Monkeys sometimes "kidnap" the babies of other females; the kidnappers are usually aunts or juveniles that are not yet mothers.[21] Like humans, monkeys allow others to approach and stare at their newborns when they are safely cradled in their arms, and they permit some cooperative care from close others, but they are also highly suspicious of untoward attention. Because of this cultural norm—in people and monkeys—we avoid interfering with a strange child if we can "help it," but the exceptions prove the rule. For example, we would not approach a lost toddler in the mall if the mother or other relevant helpers (such as a mall cop) were near or if we didn't know what to do. However, a parent may be more likely to intervene because they have experience with children and know how terrifying it is to lose a child in public, and most of us will follow the urge to help if it is clear that the child is helpless, alone, and needs us.

Our broader attraction toward neonates is somewhat linked to the "single victim effect" (aka "identified victim effect").

Behavioral economists have demonstrated many times that we respond more to requests for donations when the victim is a single individual in need over many individuals—even just two.[22] This seems irrational because surely you should want to help when there are more people in need (again returning to Bloom's idea of being "against empathy"), but when there is only one victim, the aid seems more concrete, easier to understand and contemplate, and more feasible. Beyond this, the altruistic response model predicts that individuals drive our aid *because* they resemble the canonical case of a neonate. It is probably not accidental then that many single victim experiments employ pictures of children instead of adults, suggesting some recognition that we sympathize more with younger individuals, all things being equal. Wild species conservation materials also often use pictures of cute young animals, such as baby seals or polar bear cubs to raise money.[23] In fact, in one study that we performed, people were more likely to help an adorable sea otter than a homeless man or a group of refugees. It is possible that species that deliver and care for multiples, such as rodents, cats, dogs (or humans who are genetically predisposed to multiples), are less prone to the single victim effect. To heighten the altruistic response, charities should feature young, neotenous victims who clearly need help in order to approximate the motivating schema of our own helpless offspring. Even if dozens or thousands of adults need food, clothing, or shelter, a poster depicting one cute child in clear need should elicit more aid than the huddled masses.

Given that the altruistic urge evolved to help our own helpless neonates, the urge is strongest in that context. More research is needed on the response to related versus unrelated neonates, across species, with an ecological eye toward whether the species being tested usually discriminates between related and unrelated individuals, lives in an interrelated social group, or helps more in a parental state or in the presence of hormones, onlookers, or the real parent. I expect differences in the response across species that reflect their ecological conditions; animals are more likely to care for non-offspring when they rarely encounter strange infants in the wild or, conversely, live in tight social groups, especially interrelated ones.

DISTRESS

Because *Homo sapiens* is an altricial caregiving mammal, it is often assumed that we evolved an enhanced sensitivity to cues of distress that historically helped us respond to our own helpless offspring. For example, Sarah Hrdy has written about how ape mothers are extremely sensitive to cues of distress in their young, constantly shifting the bodies of the infants clinging to their abdomens while walking to ensure they remain comfortable and attached. Hrdy also quotes primatologist Carel van Schaik, who describes orangutan mothers as responding to infant cues "with the attentiveness of a private nurse and the patience of an angel."[24]

Evolutionary neurobiologist Paul McLean wrote about how the protomammalian brain evolved into the current mammalian brain by expanding regions that help neonates utter distress calls that draw adults toward them.[25] I was lucky to meet McLean one time when I was working at the National Institute of Mental Health in Poolesville, Maryland, but my appreciation for him developed decades later, when I went back to read his many articles on the topic. It is in vogue right now to refute McLean's triune brain theory because we have discovered minor differences in the neuroanatomy and functional specialization of brain areas by species and time, and because the term "limbic system" is criticized as inaccurately implying a clearly defined circuit that includes certain brain regions, always and only. In my view, differences in the neuroanatomy across species highlight the plasticity of a brain that adapts to each ecology, but these differences do not undermine the general concept of a homology (discussed in chapter 2 on homology).

A sensitivity to distress even exists outside of primates or mammals, but it is more pronounced in altricial species. For example, mother birds are also highly motivated by the begging calls of their hungry young chicks, which prompt her to find and deliver food until the offspring can forage on their own. Some "brood parasitism" species even exploit their sensitivity to these releasing stimuli (e.g., the presence of eggs and the calls of chicks) by placing their eggs into the nests of other species. Once the strange eggs hatch, the foster chicks make particularly loud and aversive begging calls to ensure they are fed by their new foster mother. Common cuckoo chicks can even mimic the begging calls of up to eight different species to ensure that they find a suitable host.[26] The "screaming" cowbird gives more intense calls than the host's own offspring, even for similar levels of hunger.[27] This intensity ensures that limited food spread over many chicks will still find its way to their hungry mouths. These intense cues of distress must be hard to ignore if the mothers respond to them even more than to her own related offspring. This adaptation even alters the brain, such that screaming cowbirds have a larger hippocampus than similarly sized birds that are not brood parasites, presumably because the mother needs this advanced spatial memory capacity to remember where she put her various parasitic chicks.[28] The hippocampal complex in the brown-headed cowbird is also larger in females, which place and track up to forty eggs, compared to males that do not help with this process.[29] Moreover, the volume of the hippocampus in parasitic cowbirds is larger during the breeding season than in the "off season."[30] Thus, the brain is plastic and adjusts to the needs of each species, individual, and context. It would be interesting to know whether bird mothers are also affected by faces or body postures of chicks, to further determine the degree that visual versus auditory cues influence the response across species. For example, in humans, Abigail Marsh and colleagues found that faces that show fear in another person activate the amygdala and elicit approaching and helping, even when you cannot hear any cries of distress.[31]

The Cries of Newborns

The cues of an isolated rodent pup, like those described in the introduction in connection with William E. Wilsoncroft's dams, involve sounds that are attuned to the rodents' own auditory system, which are 50–70kHz higher than what humans are accustomed to hearing. Because of this, we refer to pup distress cries as "ultrasonic," even if they are perfectly "sonic" to the dams. Myron Hofer and colleagues have described many interesting features of the link between ultrasonic cries and care.[32] For example, the dam's nervous system is adaptively geared to detect these pup cries by tuning the overall boundaries of the auditory system to the frequency range of the cries while also detecting sounds at a lower threshold that sits within the range of pup cries. Pup cries also facilitate caregiving by initiating dams' active retrieval and their passive licking, grooming, and feeding of the pups. The cries even directly inhibit the rat's biting response, suppressing the tendency to bite or cannibalize the pups that they must carry in their mouths—a response that does sometimes occur in nonparental rodents.

The average fundamental frequency (F_o) of human infant cries usually ranges from 350 to 500 Hz (i.e., the vocal folds open and close 350 times per *second* or more).[33] Human infants also emit different types of cries for different problems, with the pain cry being the most intense, which attests to our evolved predisposition to focus on urgent survival issues. This range of frequencies in infant cries also corresponds to the peak frequency of sound detection in humans—approximately 200 to 500 Hz, according to the iconic Fletcher and Munson curves from 1933.[34] The top of the baby cry frequency is, perhaps not accidentally, also the sound formant frequency where singers produce the most emotionally evocative sounds, called F_4, which peaks around 3000 Hz and allows us to hear sopranos above a loud orchestra. This is known as the "the singer formant region." Research on newborn cries shows that helpless infants cry ten times as much in the hours just after birth when separated from their mothers than when together, analogous to the "separation cry" of newborn rodent pups.[35]

Indeed, the cries of a human newborn are so salient that they are used in human laboratory experiments to induce stress. One of the most common video clips that experimenters use to induce sadness in laboratory participants is the scene from the 1979 remake of *The Champ*, where a young Ricky Schroder sobs after his boxer father (played by Jon Voight) dies after a particularly grueling match. This clip even beat out another top pick by experimenters to induce sadness, where another cute young boy cries in *Kramer vs. Kramer*, a film from the same year.[36] In one study, people felt better if they ate chocolate after being saddened by the clip from *The Champ*, perhaps implicating the mesolimbo-cortical system, which participates in both offspring care and consummatory rewards.[37]

Distress cues are expected to be particularly salient for caregivers hearing the cries of their own infants. For example, when audiotapes of babies crying are played to mothers, the cry of one's own baby produces a larger heart rate and skin conductance response. In contrast, the cry of strange infants elicits an orienting response in which the heart rate decelerates and attention becomes focused on the victim.[38] Of course, this discrepancy does not mean that people help only their own offspring. In fact, the orienting response is considered a physiological correlate of empathic concern, whereby an observer carefully attends to a stranger in need and is more likely to help.[39] But for a bonded caregiver, those infant cries are not just alerting and concerning; they are also motivating signals for action. Parents also have extensive experience with their baby's cries and they can distinguish the need on the basis of the cry type, allowing them to respond not only quickly but also accurately. For example, parents know a hunger cry from an injury cry and probably respond more quickly to the latter.[40] Because primary caregivers are so attuned to their own infants' distress, they probably also detect and respond to cues of their infant in noisier environments or at longer distances. Caregivers may also find babies' cries less aversive because they are familiar with the sound, its meaning, the individual, and the response. For example, the cries of infants from parents who speak French versus German are distinguishable from as early as three days after birth.[41] In sum, cues of distress like whining, crying, and screaming are used by neonates to direct their caregiver's attention; in turn, caregivers perceive those cues as important, motivating, and hard to ignore—which promotes a response.

Due to the homology between offspring care and altruistic responding, we assume that cues of distress also elicit altruistic responses from strangers in situations such as heroic rescues. When an observer hears an adult cry from intense pain or distress, the sound is unusual and alarming, given the high pitch and volume. This intense cry also signals vulnerability and immediate need, which fuses all three features of the model into one motivating force. For example, even children in a developmental study on empathy and altruism responded to another child crying in an adjacent room, despite having to leave their assigned task to do so.[42]

The Downside of Hiding Distress

There is an important converse to the rule that people respond to clear distress: they are *unlikely* to respond when distress is missing. For example, in cultures that value independence and competence, it is considered a sign of weakness or immaturity to express distress through crying, screaming, or whining—especially for males and outside of childhood.[43] In such cultures, even young children who cry from a painful experience are advised to "stay tough" and are overtly discouraged from crying. This quelling itself reflects the action of the offspring care mechanism because the salience and aversiveness of the sound is so associated with infant vulnerability that people want it to end as soon as possible. Because of these dynamics, which are exacerbated in some cultures, people can hide their distress to avoid being viewed negatively.

The fact that people learn to "swallow their pain" also means that they often fail to receive aid when they really need it. In my Ecological Neuroscience Laboratory, we videotaped interviews with real hospital patients who had a serious or terminal illness.[44] When our participant observers watched the videotapes of the patients, they were more likely to report that the highly distressed women needed the most aid, they felt more empathy for them, and they offered them more help. In contrast, the reticent man who did not show emotion did not elicit empathy or aid, even though objectively his illness was just as severe. Like our reticent patient, a follow-up study in another lab that videotaped interviews with college students found that observers were less empathic toward students who did not express emotion when describing their personal problem.[45] We evolved to respond to clear distress and, so, when people "save face" by stoically masking their feelings, we are precluded from realizing their plight and wanting to help.

Observers often have the knowledge that an unexpressive victim needs help but still fail to respond because of the inherent link in the offspring care system between overt distress and neonates in need. For example, you might know that your friend has a scary doctor's appointment or an important test the next day and send a comforting call or text, even if they do not display distress to you. This effortful empathy occurs less often and is appreciated by the recipients because people so rarely make this small effort to imagine how another feels. Paradoxically, suffering individuals underestimate how hard it is for others to infer and respond to pain that they inhibit and end up sitting in silence, which makes them resentful on top of being in pain. People do put forth the effort to imagine another's suffering when it is a close friend or family member whom they care for and are interdependent with, and even when people are unliked but share in our fate, such as an irascible boss that we must keep happy. But most of human helping is not well explained by the view that we take people's perspective, because the task is effortful and produces only a weak motivation to act compared to the altruistic urge. Perspective taking cannot explain why people as well as nonhuman animals including rodents are motivated to care for neonates and bonded partners who exhibit clear distress.[46]

I acknowledge that sometimes it is important to save face. For example, because adults appear more subordinate and

incompetent when they display infantlike distress and vulnerability, an employee may not want to cry in front of a colleague
or boss, particularly women who are already stereotyped and
denigrated as being too emotional. For example, on the 2008
campaign trail, there was a brief moment when a fatigued Hillary
Clinton looked misty while describing her passion to help
America along with the demands of campaigning:

> It's not easy. It's not easy. And I couldn't do it if I just didn't, you
> know, passionately believe it was the right thing to do. You know,
> I have so many opportunities from this country. I just don't want
> to see us fall backwards. You know, this is very personal for me.
> It's not just political. It's not just public. I see what's happening.
> We have to reverse it.
>
> Some of us just put ourselves out there and do this against some
> pretty difficult odds. And we do it, each one of us, because we
> care about our country.
>
> But some of us are right and some of us are wrong. Some of
> us are ready and some of us are not. Some of us know what we
> will do on Day One and some of us haven't really thought that
> through enough. . . .
>
> So, as tired as I am—and I am—and as difficult as it is to kind
> of keep up what I try to do on the road, like occasionally exercise
> and try to eat right, it's tough when the easiest food is pizza. I
> just believe so strongly in who we are as a nation.

Her display of vulnerability lasted mere seconds, and yet it caused
an outpouring of media attention that cast her as either staging
emotion to seem more human (because she is perceived as mas-
culinized) or as proving the negative stereotype of women (that
they are too emotional to be in charge).[47] Damned if you do,
damned if you don't.

As such, there are times when you need to reveal your pain, even in the face of possible disregard, and times when revealing your vulnerability is too problematic, particularly for individuals striving for respect in the face of considerable bias. There are also times when displaying your distress could be problematic, and may even cause problems, but you should consider displaying it anyway. For example, an irascible boss may not realize how hurtful and mean he sounds and might regret his actions if he could clearly observe the effect of his words and actions on his valued employees. He might even change. At the very least, if you clearly exhibit distress when he hurts you, he cannot claim ignorance later when you launch a complaint. Even when we feel compelled to hide distress, we might sometimes benefit from just letting this powerful signal do its job.

It is even more problematic when we hide distress from close partners who *want* to help us and are less likely to take advantage of our vulnerability. For example, during my first pregnancy, I did not tell anyone that I was pregnant for the first three months, subscribing to a cultural rule to avoid having to "untell" people if the pregnancy were not successful. I was explicitly discouraged from laying bare my suffering, because it was assumed that revealing a miscarriage was embarrassing and would only augment the pain of an already heartbreaking event. But then a nurse said to me, "Well, if something bad did happen, aren't there people that you would *want* to know so that they can support you in a difficult time?" This *benefit* of alerting your friends and family to a problem was an important message to me about the negative consequences of playing it close to the vest.

People also often discover too late that someone they cared about suffered from a mental illness like depression, anxiety, or an eating disorder. People with these conditions often mask their symptoms or need because they feel ashamed or weak, or worry about being judged or pitied. People often say when someone in their social circle takes his or her own life, "I didn't even know they were depressed!" Sometimes even close family members are surprised. The fact that we discourage openness about crushing mental and physical health needs—even within our own families—has dire consequences that are avoidable. We must work harder to normalize the expression of suffering, particularly among close others that we want to help.

In sum, distress evolved to be a cue that is strongly associated with vulnerability and need, which motivates observers to act. But because it is associated with vulnerability, and has such a powerful effect on observers, people often hide their distress even when they really do need help. Sometimes this concealment is beneficial to avoid the associations between distress and weakness or vulnerability, but too often it disadvantages the victim and observer alike. Emotions evolved for a reason. Sometimes we need to remember that emotions are powerful tools that can help us in the long run if we let them do their job.

The Psychology of Distress, Empathy, and Altruism

We argue throughout that distress evolved to be salient, demand our attention, and compel action in the context of offspring care. This tenant of the altruistic response model appears to conflict with the widespread view of C. Daniel Batson and Jean Decety and others that distress impedes aid. According to the empathy-altruism hypothesis, people focus on others' needs and offer selfless aid when they feel warm, tender, calm, concerned and compassionate; in contrast, they focus on their own needs and help only if alleviates their own distress if they feel worried, concerned, distressed, perturbed, and upset.[48] As an example, when students in a laboratory witnessed someone receiving painful electrical shocks, observers who reported feeling empathic concern helped even if they could leave whereas those who felt personal distress helped less, unless they were forced to stay and continue watching the painful shocks. Thus, people are *capable* of helping for selfless reasons, but can also act selfishly to assuage their own distress.

Our own research does sometimes reveal problematic forms of distress. For example, we usually replicate the finding from Batson that distressed victims can evoke sympathetic as well as negative responses in observers. When people view the video-tapes of our most distressed hospital patients, a subset of par-ticipants even report feeling *horrified* (i.e., perturbed, angered, horrified). This highly negative response is all the more strik-ing since participants know that these are real patients with serious or terminal illnesses. Thus, aversive responses *can* occur to another's expressed distress when it leads to unwanted, contagious, negative feelings—particularly when their issue seems unwarranted or difficult to resolve. (For example one nurse commented, "Well, what is she doing about this prob-lem?") All is not lost, however, since the average person actually sees more need, feels more empathy, and offers more help to the distressed patients over the happy ones. This spirit of generosity does have its limits, though. For example, participants increase their relative preference to assist the happy patient if they have to sit with the patient instead of just giving them a few dollars without any social contact.[49] Thus, even though distress is defi-nitely aversive to perceive and feel, it does successfully convey need and stimulate a response, just as it was designed to do.

We can predict these complex relationships between distress and altruism if we contemplate how the attributes of the altruistic response model trade-off in any given situation. For example, people complain when babies cry on the airplane. This seems paradoxical, since those are actual babies in distress, which we supposedly evolved to help. However, this irritation fits the model because those babies are not familiar or bonded to the rest of the passengers, most of whom are too far away to be enraptured by the baby's cuteness and do not know what the problem is and cannot help. Thus, the classic "baby crying on an airplane" is naturally distressing—which proves that the sound is salient and motivates us to stop it—but we fail to empathize or help because we lack the bonding, familiarity, expertise, and control that define parental care, and we are further inhibited by social norms to leave strangers' babies well enough alone. This conflict is all the worse in cases like child abuse, in which caregivers withdraw or even aggress against or injure distressed children that they are supposed to protect. According to research, because distress is so salient, motivating, and impossible to ignore, people become overwrought when the distress or crying persists for hours or days on end, especially when there is no clear solution (e.g., because the baby has colic).[50] People must be trained and supported in these situations and not chastised; they should be able to remove themselves from an intense situation in order to calm down and we need to provide help so that the caregiver can have a break. This situation is contributed to by a disconnect between the supportive, social, group life that we evolved to raise children in and the Western, industrial condition of solitary parenting that most of us experience today.

In contrast to these cases in which distress does not promote aid, even intense and aversive distress can still yield aid if the observer understands the situation, can intervene, and feels confident in their response.[51] The neurohormonal stress response in mammals did not evolve so that we could eat cookies when we are stressed about work, it evolved to promote immediate action by marshaling sympathetic autonomic and metabolic processes at the expense of slower, longer-term bodily processes like digestion and growth.[52] Our stress system evolved to maximize the efficiency of a fast response under duress, such as when a stressed observer must leap into action to help someone—assuming they know where to jump and how high. Thus, even if distress cues activate your stress and autonomic nervous systems, they can produce apathy, irritation, or aggression in situations where we *cannot* act—when there is no clear outlet for our intense arousal and distraction—because these states evolved per se to promote action.

People can also be conflicted when they face distress if they think they might be manipulated. Because distress promotes aid, people sometimes fake distress to elicit support, which can embarrass, irritate, anger, or disgust observers who become suspicious of the victim. For example, Sarah Hrdy has described how New World monkeys that breed cooperatively like marmosets and tamarins usually share food with helpless babies, particularly when they beg for it. However, as juveniles age and gain independence, adults are *less* likely to share food with them, which causes the juveniles to beg in increasingly intense and aversive ways to solicit the food, sometimes resorting to stealing it. This phenomenon has been replicated in the famous vampire bat animal model of altruism, wherein adults share less of their blood meal with bats that have developed past the juvenile stage and should be providing for themselves.[53]

Babies are genuinely helpless and, at least in the early phases of infancy, cannot really be considered to "manipulate" their caregivers with cries, at least not in the intentional or diabolical way that toddlers, older children, and adults might. Infants may "use" cries to inspire their caregivers to give them food, warmth, comfort, or to remove a noxious stimulant. It is one of their only ways to communicate need. Some of these needs are not true emergencies, but even the need for passive care, such as physical comfort, can affect infants' long-term health and well-being. For example, babies often cry when left alone in a crib or car seat because they prefer the warm and loving embrace of their caregiver. But these are not acute needs that require an immediate solution (and in the case of the car seat may be the very thing that saves them). Even though babies use cries to inspire us to help them, I think we can agree that they are not plotting against anyone intentionally and have pretty reasonable requests—especially in the face of some pretty annoying modern contraptions. Thus, the genuine, unexaggerated, unmanipulated quality of a baby's cry provides a releasing stimulus for action, which we still abide when issued by an adult.

People are sensitive to the qualities of distress cries, and can distinguish cries that reflect different needs such as for contact, hunger, and pain. As such, the newborn who gets a shot in the leg at the doctor's office elicits more sympathy than the eighteen-month-old at the library who is half whining and half crying about a toy they cannot take home. The latter form of whining and crying can be very annoying to observers, who may even get upset with a child that they view as manipulative, particularly when the goal is to acquire a reward like a train or more Goldfish crackers. However, people *do* feel for the true neonatal cry of need, which is softer, patterned, and suggests the ideal combination of being vulnerable, neotenous, distress, and in need.

Distress is not one thing. It comes in many forms and contexts, some of which are motivating and some of which are not. But if we understand distress from the context of caring for a helpless neonate, patterns emerge. A genuine distress that is caused by a serious and immediate situation that requires help that the observer can provide is motivating, whereas it is aversive and less likely to promote aid when the observer is not familiar or bonded, does not know what to do, is not in a position to help, or feels manipulated.

The scientific literature needs to be more specific about when distress causes people to run toward versus away from difficult situations, and to contrast situations that do or do not entail attributes of the altruistic response model, such as whether the victim is bonded to the observer, is neotenous, displays distress overtly, and needs immediate help that the observer can provide. These studies would provide a more complete picture of the range of responses to distress in the real world, which is not always compassionate but does yield many more possible outcomes than just self-focus.

SUMMARY

The altruistic response model proposes that individuals are inspired to help—feel an *urge* to help—in situations that mimic the need of helpless offspring. As such, features that are intrinsic to infants are designed to inspire our urge to respond, even when the victim is an adult or stranger. Babies, by definition, are young and vulnerable and need aid that we can provide. Sometimes adults also possess some of these features, which additively combine to promote our urge to respond. Vulnerability, distress, and immediate need are probably stronger signals than neoteny, all things being equal, but they are also responsible for our failures to act, such as when we are unmotivated by chronic need, hidden distress, or issues we cannot observe directly. These victim attributes do not operate in isolation, like on-off switches. They are integrated implicitly and quickly by the brain, through normal, dynamic information processes that aim to produce the most beneficial response for each situation, given what we know about it. It is important to understand how the altruistic response occurs in evolution and in our bodies so that we can predict when responses will occur and counter our regrettable biases to sit idly by.

7

CHARACTERISTICS OF THE OBSERVER THAT FACILITATE A RESPONSE

I n the case of William E. Wilsoncroft's assiduous dams, pups naturally possess all of the appealing features of victims in tandem: they are vulnerable, helpless, and need aid; moreover, the dams know how to help . . . and so they do! The same goes for humans. Assuming that the victim possesses such infantlike characteristics, there are also features of the observer that can inhibit or enhance the likelihood that the altruistic urge will take place. I described victim characteristics first because, all things being equal, they would appear to influence the response more, given that people are compelled to help victims who truly resemble helpless neonates, whether they are naturally empathic or not. There is an exception to this rule of thumb, though: even a "perfect" victim will not be helped unless the observer knows the appropriate response and predicts that it will work, in time. This is a unique focus of the altruistic response model, relative to other theories of empathy or altruism, because it construes altruism as an *act*. Actions are controlled by the motor system, which is designed to implicitly predict how our behavior is going to work, and how others will respond, so that we can quickly and adaptively adjust to unfolding situations. There are four relevant attributes for observers that inform whether they will respond to a victim:

- Expertise
- Self-efficacy
- Presence of observers
- Observer personality

EXPERTISE

Expertise is described first because it is one of the most important and unique attributes of the altruistic response model. According to the model, people respond to victims when they can predict—even implicitly and outside of conscious awareness— that they know and can enact the appropriate response in time.

This focus on expertise is related to the fact that the neural circuit for retrieving offspring involves two opposing circuits—to avoid versus approach the pup.[1] This opponency naturally prevents individuals from helping when it would be maladaptive, for example, when a novel stimulus seems dangerous. Rodents thus avoid retrieving pups when they are not primed with the appropriate hormones or habituated to pups but they approach and retrieve pups when they are primed and feel comfortable. Humans similarly avoid helping when they fear novelty, uncertainty, or danger but approach when they feel prepared and in control.

Expertise is relevant to the altruistic response because helping usually requires an overt motor act. Spontaneous acts such as pulling a child back from danger could rely upon fairly simple motor programs, which may even be innate or part of basic human motor capacities. Thus, simple retrievals may not require motor expertise and could be issued by any observer—but only if the observer predicts that they are fast and strong enough to pull the victim back from harm. Sometimes retrievals require *exceptional* strength, skill, or knowledge, usually those are the acts that we construe as truly heroic. Emergency situations unfold quickly, and the observer has little time to confidently select the best response right when it matters most. Human altruism is also required in a wide variety of contexts beyond simple retrievals, such as leaping into rushing waters or locating someone trapped in a burning or collapsed building.

Some existing research supports the benefits of expertise. For example, sometimes apathetic human bystanders later report that they failed to act because did not know "what to do."[2] Similarly, firefighters describe a learned but intuitive understanding of how safe or risky a house fire is in any given moment, allowing them to make quick, safe decisions in a situation that is constantly changing.[3] The aforementioned case of Wesley Autrey, who jumped onto subway tracks to save the young man who fell on them after a seizure, also possessed rare and relevant expertise.[4] Wesley credited his fast, life-saving response to his extensive prior work as a construction worker who worked in confined spaces, which allowed him to predict accurately that the two of them could fit under the oncoming car. Similarly, Good Samaritans who intervened in crimes were found to be larger, with more training with crimes and emergencies, and they described themselves as "strong, aggressive, principled, and emotional." Their inspiration to act appeared to emanate from "a sense of capability founded on training experiences and rooted in personal strength."[5]

Thus, just as a soccer player can implicitly and quickly calculate how hard to kick the ball by integrating the distance to the goal, the position of the nearest defender, and the condition of an old ankle injury, an observer can implicitly and quickly calculate whether he or she can reach and rescue a victim in time. Taking Wesley Autrey as an example again, he would have had to calculate how far away the train was, how much time he had, how long it would take to pull the young man out, and whether he could lift him out by himself, quickly. Wesley predicted that he could *not* remove the victim in time, but that they could fit under the car if he lay on top of the young man. Thankfully, his prediction was correct, and the train passed narrowly over them, allowing them both to survive. Wesley made a complex calculation almost instantaneously that came down to a matter of centimeters, with dire consequences if he was wrong. Any one of us would make some calculation in that moment, but his expertise surely improved the accuracy of his guess, and the fact that both are still alive today.

The motor system in the brain is designed to make quick, accurate predictions about appropriate responses in time. For example, there is a classic finding in the motor literature that people can accurately guess whether they can walk up a flight of stairs just by looking at steps of varying heights; once the riser surpasses a certain height, people uniformly switch from guessing yes to no.[6] Primate brains also predict which motor act is needed to pick up various objects. The exact location in the brain and which cells are engaged shifts within the premotor cortex depending on the appropriate grip needed to grasp an object, for example, using the "precision grip" to pick up a raisin versus a "whole hand grasp" to retrieve a jar.[7] Applied to altruistic responding, people are good at quickly guessing whether they can retrieve a victim in time or are fast or strong enough to pull him or her from danger. Expertise is usually only discussed with respect to purely intellectual or motor skills such as knowing calculus, playing the piano, or sinking a basketball, but it is also critical for determining when we respond to someone's need.

Importantly, expertise *also* plays a role in social and emotional or passive forms of altruism, such as when you must console a distressed colleague or friend. When faced with a distraught person—or even thinking about someone who might get upset—people implicitly (and sometimes explicitly) predict what will happen if they try to intervene: Will my friend become more or less distressed if I ask him about the issue? If my friend gets more upset, will I be able to help him feel better? When you know the victim, you can better anticipate that person's needs and how to respond—a form of expertise that developed over your lifetime of social interactions in general and with that person. You also have expertise with your own prior attempts at comforting others, which increases or decreases your faith that you can help. People from families or cultures where open displays of emotion are discouraged not only show less of their own distress, but they also feel more uncertain in the face of another's distress, having had little practice with these dynamic and intense situations. Both parties then suffer in silence, wishing for contact, but unable to make the leap because they lack the experience to understand how to use and modify these intense emotional states.

People's implicit or explicit prediction that they will succeed—in both physical and emotional situations—strongly influence the likelihood of a response. Expertise overlaps with the concept of self-efficacy, which can influence behavior more broadly.

SELF-EFFICACY

"Self-efficacy" refers to the fact that people do not act when they feel that their efforts will not achieve the desired outcomes.[8] Self-efficacy influences behavior in a variety of contexts, from education to work to social behavior to recycling. For example, when people construe recycling or carpooling as requiring them to relinquish valued independence and comfort, as confusing or difficult, or as making a tiny dent in the huge problem of climate change, they just don't bother.[9] Applied to altruism, a lack of self-efficacy makes people apathetic when they don't think they can improve the situation. This overlaps with expertise, because both involve a prediction about the results of one's actions; however, self-efficacy is also implicated when people can *physically* enact the response, but do not assume that their aid will ameliorate the problem. For example, when a homeless person asks for spare change or a charity needs money to feed a starving nation, the donation itself does not require motor skills, and you might even have enough to give. However, the problem may seem too large or complex to solve with a few dollars, which throws a bucket of ice water atop any burning motivation to help.

In the theory of planned behavior (figure 7.1), motivation translates into action only if the person—as well as people in their valued social circle—value the act and also believe that they have *control over the outcome*.[10] Thus, even if you think that recycling or helping the homeless is the right thing to do, you won't form the intention to carry your used office paper down the hall or to retrieve a dollar for the man on your block if coworkers or friends say that recycling is all landfilled anyway and homeless people just use money for drugs.[11] The element of being able to control the outcome is similar to expertise and self-efficacy, because in all of these cases you must predict that your act will alter the outcome for the better. The same goes for altruism.

Most people believe that altruism, in general, is a good thing (i.e., they have a positive attitude and social norm); however, they vary greatly in how much they believe that their aid will do any

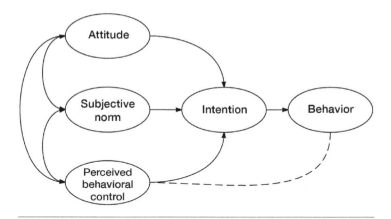

FIGURE 7.1 A diagram of the theory of planned behavior, which importantly presumes that people must value an outcome to perform a desired act and view the act as appreciated by important others and view it as feasible and likely to succeed.

Drawing by Robert Orzanna, CC-BY-SA-4.0, based on theory from Icek Ajzen, "The Theory of Planned Behavior," *Organizational Behavior and Human Decision Processes* 50, no. 2 (December 1991): 179–211, https://doi.org/10.1016/0749-5978(91)90020-T.

good, such as giving money to aid the poor, help an addict, or save a tribe of refugees. Some people are less sympathetic and offer less aid because they believe in a "meritocracy," such that all success is viewed as resulting from hard work (without noting any help they received along the way); conversely, they assume that if people are not successful, then they must not be working hard. Others are more sympathetic and support assistance because they consider how situations such as a safe loving home, high-quality local schools, and high-status peers and mentors are needed to promote success in conditions of poverty, abuse, or illness—regardless of how hard one works. Thus, just as a hero must believe that he can swim out to the drowning kayaker in time, the politician must believe that diverting tax money into public assistance can pull people out from debilitating poverty and onto more equal footing.

Self-efficacy also undermines action when people perceive the scope of a problem as just too big to be solved.[12] For example, when it comes to "saving the Earth," people are overwhelmed by the amount of waste and assume that only corporations can do anything about the problem, feeling helpless to effect any real change. The scope of the problem also prevents action because, for example, you might feel as bad as you can when you learn that ten people are suffering; this concern cannot scale up to meet the size of the problem when it turns out that ten thousand people are suffering ("scope insensitivity").[13] This problem of scope interacts with the single-victim effect described earlier, in which people prefer to help one person over two or more. When you give money to a specific person, even a small donation that you can afford, it seems possible that your gift could alter their day or week, which adds to your sense of control, efficacy, and your "warm glow" from giving. This is quite in contrast to the depressing feeling that people get when they consider how their drop in the bucket will assist thousands of people, whom they cannot even see.[14] It's hard to imagine that your small infusion of aid can solve a massive problem such as a food shortage, ethnic cleansing, or violent conflict in a faraway land. And so people do nothing. Moreover, people actively avoid information about such events that they feel so helpless to change.

The Environmental Defense Fund (EDF) found a surprisingly high response by their members to a campaign to help the monarch butterfly.[15] We examined stories that people wrote on the EDF website about their experiences with monarchs. People were inspired by their view of the species as beautiful and as having a special life history that they directly observed in the garden or classroom. People knew in advance how many acres of milkweed would be planted per amount they donated, making it easy to imagine concrete success from even twenty dollars. Similarly, in another study, people gave more to our hypothetical environmental appeals when the advertisement described the earth as vulnerable, paired with the image of a single polar bear and cub afloat on a small ice flow.[16] People feel more self-efficacy and satisfaction from gifts when they can perceive the victim and the aid. Thus, larger problems motivate people *less* because it reduces their self-efficacy and their belief in making a difference.

Part of scope insensitivity comes from a perceptual mismatch between information and how our brains evolved to process it. Our altruistic response developed to assist one's own bonded offspring who need our help right now—a single, concrete victim who is directly observed. Donations to refugees in a distant country require a more difficult ability to imagine how hundreds or thousands of small donations from around the world can aggregate into a useful amount, to assist with a problem that you cannot see or understand. Even though the majority of charitable donations do come from individual donors, whose small gifts add up (compared to a few large gifts from wealthy donors), people often fail to give because they assume that their small gift cannot add up to anything meaningful.[17] This is the rub of self-efficacy. In a system that evolved to be driven by immediacy and concreteness, the bigger the problem, the *less* we want to help.

Some believe that this error in our giving ways should be addressed by skipping empathy, and instead making altruistic decisions on the basis of rational calculations of the costs and benefits of giving.[18] This calculus may work for long-ranging policy decisions but cannot sway the sympathies of individuals, since it is needed to elicit charitable donations and to garner constituent support for an altruistic policy. The features of our ingrained altruistic response are so powerful that we are better off not trying to work against them and will have more success if we understand and use the mechanism to evoke empathy and aid, even in situations that naturally do not release an urge to help. For example, some charities match givers with specific victims, such as allowing donors to adopt a single child, family, or village, or letting them furnish a microloan to a specific group of women in Africa or to buy a cow or goat for a designated village. In our research, we have increased people's donations for the environment by appealing to their sympathy for vulnerable, distressed, targets in immediate need such as a baby polar bear floating out to sea and by appealing to their attraction toward beautiful, attractive targets, like fresh mountain streams.

PRESENCE OF OBSERVERS

The altruistic response model assumes that people feel an urge to respond to victims like helpless neonates who need immediate aid that they can provide. These entailments intersect significantly with those of bystander apathy, a phenomenon in psychology in which people help less as the number of other observers increases.[19] Bystander apathy is assumed to be an inherent part of our tendency to avoid helping when uncertain or intimidated, which increases with the number of observers. Particularly in the medical emergencies that are often used in bystander apathy experiments, people can feel overwhelmed or uncertain about what is wrong with the victim, what type of help they need, if they are the best person for the job, or if they will be judged if they intervene. Expertise such as medical training increases someone's likelihood of responding because it reduces this uncertainty—mediated by the avoidance arm of the offspring care circuit—and increases one's confidence in a successful response—mediated by the approach arm of the brain circuit. In contrast, we would really regret it if we rushed in "guns blazing" if it turned out that the victim was just screaming with glee or if we ended up paralyzing the victim when pulling him or her from a crash. Such fears and uncertainties plague the minds of observers and foment nonresponse, an effect that is generally adaptive and explainable through the altruistic response model.

Regardless of the number of observers, the altruistic response model specifies that people will respond more when the victim is unquestionably vulnerable, distressed, and needing urgent aid. But it is a nontrivial task to determine if someone is truly vulnerable. Some victims are clearly dissimilar to a helpless neonate (e.g., a manipulative coworker who wants you to empty his recycling bin), and some are clearly similar (e.g., an abandoned baby on a church doorstep, in the cold). But an infinite number of cases sit in the gray area between such extremes, forcing the observer to decipher whether the target is truly vulnerable. Added to this, we rarely face adults or infants in dire need of help because adults are by their nature capable and are socialized to seem capable even when they are not, whereas babies are so helpless that they are rarely unattended by their loving caregivers.

People lean toward intervening when they are more extroverted, risk-seeking, or unconcerned about what others think of them—attributes that occur more for observers in their own neighborhood when they feel comfortable and responsible for the place.[20] Even people with psychopathic tendencies, who are by definition not empathic, help more in emergencies because they have no qualms about the danger or people's perception of them and they may even seek the narcissistic rewards of heroism.[21] Conversely, people who are risk-averse, see danger everywhere, are easily embarrassed, or have social anxiety are less likely to rush in because they fear making a mistake. People are probably more likely to help a stranger when they come from an independent culture like America over an interdependent one like East Asia.[22] They are also more likely to help when they are surrounded by people who seem supportive, such as friends, familiar individuals, and female observers (women and men help more when the observers are female).[23] Both humans and rodents help less in the presence of a stranger, which increases the stress hormone cortisol or corticosterone and inhibits helping; this effect can be reversed if you block the stress hormone or allow the stranger to become more familiar through a short, fun initial game.[24] Thus, observers help when they feel comfortable in their surroundings, are not worried about what others will think, and are not stressed by strangers, supported by the fundamental opposition between avoiding and approaching in the offspring care mechanism.

OBSERVER PERSONALITY

Many developmental, biological, and affective processes conspire to influence when an observer will help—in general, and in a particular situation. These processes typically reflect a sensible combination of nature *and* nurture, with an emphasis on how one's early environment alters genes to prepare individuals for the environment into which they are born.

In general, there is some evidence that almost any personal characteristic or life experience that you imagine *should* increase altruism does, in fact, contribute. But many of these attributes only have a minor effect on behavior, which is often overridden by the situation—whether the observer knows how to help in that context or bystanders are present, for instance. Moreover, people who perform heroic rescues often report having no idea why they rushed in and do not refer back to the way that their mother or priest taught them the Golden Rule (but people do for types of aid that are more sustained, for example, hiding Jews from the Nazis during the Holocaust). Heroes also do not report feeling the empathic concern that is most often mentioned by psychologists as the reason that people help (e.g., a calm, attentive, warm, tender state).

The Empathic Personality

When people wonder why someone might be altruistic, they most often refer to personality—assuming that some people are just naturally more empathic and altruistic than others (that they possess "trait prosociality").[25] Empathic traits are usually measured on self-report questionnaires that divide personality into attributes such as the tendency to catch other's emotions (emotional contagion), to sympathize with people in need (empathic concern), to become distressed in the face of others' distress or in emergencies (personal distress), to imagine what is it like in the other's shoes (perspective taking) or to help others with small daily acts of altruism (daily altruism).[26] Attesting to the prevalence of this perspective in academic research, a recent Google Scholar search found more than sixty thousand articles that cited the most popular measure of empathy, Mark Davis's Interpersonal Reactivity Index (IRI), with almost five thousand direct citations of his original paper from 1983.[27] I also use the IRI in my research and usually can find a correlation between the personality trait and the task that assesses empathic or altruistic behavior; however the patterns of results are hard to predict and interpret. For example, usually if one subscale predicts aid, they all do; sometimes both empathic concern and personal distress increase giving, or one subscale predicts one task but another subscale predicts the other task. The only clear pattern we have noticed is that imaginative fantasy less often predicts responses than the other subtypes of empathy. More work is needed to hammer out these discrepancies, but it is clear that observers do enter situations with a predisposition to attend to, feel with, and respond to others' need.

Much of this research into the empathic or altruistic personality is employed to support the empathy-altruism hypothesis,[28] in order to demonstrate that people help more when they possess a trait tendency toward empathic concern compared to personal distress. In reality, the sheer magnitude of the impact of these personality measures are often small, empathic concern and personal distress are often interrelated, and sometimes personal distress predicts aid. Too often these traits do not predict people's actual response to another individual. Specific to the altruistic responses that are the focus of this book, experiments rarely test how empathic concern or personal distress affect peoples' actual behavioral responses to someone in urgent need. The daily altruism subscale in Lou Penner's Prosocial Battery does ask about acts of altruism,[29] but for innocuous aid, such as giving someone directions or the time, acts that are less likely to be affected by the type of victim or the expertise of the observer. Thus, personality trait scales of empathy or altruism are important but imprecise measures, which do sometimes statistically predict giving but are not designed to predict how people will respond with an overt act in a specific, difficult situation in the real world.[30]

Observers also differ in their likelihood of attending to the other person in the first place, and this attention is required for the brain to process and appreciate a victim's state. Attention to others requires that you are interested in them or their general welfare in the first place, believe that their state is relevant to your goals, and do not expect that being involved will bring you problems such as contagious distress or a conflict with needing to help yourself. This attention to others is partly explained by your personality, but also with many other factors, including your ability to regulate emotion, emotional attachment to others, concern about strangers, and even your larger worldview (is life "brutish and short," or are we all in this together?). Even after attending to the victim, you must still determine whether to respond, which again is affected by many factors: Is the victim "victimlike"? Are you responsible? Are you affected by their emotion, interdependent with them, or stressed? All of these factors and more contribute to what we consider a prosocial personality. A "prosocial personality" sounds like one thing, but it actually reflects the combined influence of many lower-level developmental, psychological, biological, and epigenetic processes that are separable but that come together in a systematic way that allows us to measure their combined influence over time, within a person. Thus, a questionnaire about how prosocial you are (or want to seem) can inform how you see yourself relative to how other people see themselves, but not *how* or *why* you came to be that way—questions that must be answered through lower-level neurobiological and psychological factors.

The Positivity Bias

Our view of ourselves and even our memories are biased by the way that we tend to focus on our good attributes, what was most memorable, what we wish to forget, and so forth. For example, if you ask me if I am a "clean" person, all I need to do is remember that I washed my hands ten minutes ago and asked my kids to vacuum this weekend and I can say, "Sure! I'm pretty darn clean."[31] In that moment, I do not consider how clean I am *relative to* other people, or how many times I could have cleaned and didn't. Applied to altruism, if you ask me if I am helpful, I can surely retrieve a salient memory of a time when I was helpful of which I am quite proud and confidently affirm the positive. Women may also report more empathy than men because they are socialized to seem helpful. (Some scholars find similar levels of altruism between men and women, even if women report more empathy, but I always find more empathy and more giving in women.)[32] Researchers attempt to factor out such biases to want to seem or look good (which they call "social desirability") by administering a separate questionnaire that measures social desirability and then subtracting that influence out when predicting an outcome from a personality variable like empathic concern. This approach, however, does not work so well for a trait like empathy, which is inherently tied to people wanting to be nice or to do good. People are also notoriously bad at reporting on their own motivations.[33] Even if we value good acts and could accurately reflect upon them in surveys, the theory of planned behavior predicts that people will still fail to help when valued peers do not share the value in helping or when we might not be able to change the situation. Thus, rather than ask people if they approve of recycling or want to be helpful, we should ask people whether they plan to change a lightbulb from incandescent to CFC or to donate to a charity *this week*.[34] Taken together, personality questionnaires do measure real differences across people, but reflect many interacting underlying processes; they cannot predict whether someone will perform an overt rescue in a particular situation the way that the altruistic response model can.

Early Development

When you examine the factors that promote prosocial behavior in children,[35] researchers find that people are generally more helpful when they are conscientious, want to please others or ingratiate themselves with valued social partners, and are raised by authoritative parents who value being a good person and consider how their actions affect others. Parents who not only explain the "right" thing to do but also model compassionate, caring, and warm behavior raise children who are more prosocial than parents who are harsh, unyielding, or cruel. Similar effects emerge in animal models, in which attentive rat or monkey caregivers produce offspring that are also more social and attentive, whereas inattentive or cruel caregivers raise offspring that become poor parents, anxious adults, or aggressive social partners.[36]

Psychopathy

Parents who are overly harsh, critical, unsupportive, or physically or emotionally abusive can raise children that are more likely to become sociopathic or psychopathic.[37] A child raised in an abusive or unsupportive environment develops an inability to feel for the distress or pain of others in an emotional and compassionate way.[38] Robert James Blair and Abigail Marsh and colleagues have demonstrated that individuals with psychopathy are impaired in their emotional response to others' distress and fear, linked to a smaller sized amygdala and lower brain activity in this region—the one that we described earlier as being needed to switch the offspring care system from avoiding to approaching pups.[39] In contrast, "extreme altruists" are *more* responsive to others' distress and fear and have larger amygdalae and stronger brain responses in this region.[40] Importantly, such extreme altruists do not just report being helpful on surveys or artificial tasks; by definition, these individuals have made heroic sacrifices for others in real life, like donating a kidney to a perfect stranger.

Sometimes people believe that people with psychopathic tendencies are even *better* at perspective taking than the average person, which helps them to manipulate others. Psychopathy can increase one's desire to rush into a public emergency; however, the ingenious psychopath from crime shows such as *Criminal Minds* is a rare bird who perhaps exists somewhere but is almost never seen. Most individuals with psychopathic tendencies are raised in difficult conditions and have trouble sustaining normal, productive adult lives—often ending up incarcerated as a result of their risky, impulsive, and aggressive acts. These conditions do not really attest to the ingenious planning capabilities or the expertise in human behavior that people attribute to psychopaths in the media. A study that examined psychopathy and intelligence in a database of prisoners found that prisoners with higher IQs performed more violent crimes, earlier, than prisoners with middle- or low-range intelligence.[41] Thus, even in an intelligent and dangerous form, psychopaths may be manipulative and harmful but are not mad geniuses who outwit us at every turn. Despite this, traits such as being callous, unemotional, and self-serving can lead to success when they are limited and channeled diligently into socially accepted activities such as sales or arbitrage.[42] You can also imagine how others feel without "true empathy"—for example, if you are conditioned or taught explicitly to learn the link between your actions and their consequences.[43] As the saying goes, "There is more than one way to skin a cat."

Spirituality

Some research suggests that traditional religion or spirituality can make observers more altruistic—particularly males.[44] But those effects could reflect religious teachings or something else that drives someone toward religion in the first place, like a personal striving to be a better person or a desire to connect with people in the community. A social norm of service to the poor is also high in religious communities, which promotes an intention to act to help others. Currently, this literature is inchoate. Some studies find that religion does increase altruism, some find that it does not, and some even find that it *reduces* helping.[45] Regardless, social structures such as a bonded, spiritual community that values service are expected to drive people to help.

SUMMARY

In some ways, attributes of the situation and victim are more fundamental to the altruistic response, because victims that resemble truly helpless neonates precipitate a response in most observers—as long as the observer knows what to do and thinks it will work. Most attributes of the observer simply shift people's *threshold* for classifying a situation or victim as important (e.g., Is the person really helpless? Can I fix the situation?). To the extent that someone has a systematic bias to view situations one way or another (e.g., confident or unsure; risk-seeking or risk-averse), his or her response tendencies will be biased in the same way. You can average all of these lower-level processes into what seems like one "altruistic personality," but this variable does not reflect one underlying cause or gene but rather the summed combination of genes, early environment, upbringing, culture, beliefs, and personal goals.

The most profound observer attribute in the altruistic response model is a belief that one can succeed in helping. This implicit or explicit prediction is embedded in our motor expertise for heroic acts but includes "self-efficacy" for more common types of aid that we furnish when we believe our gift will make a difference. As such, altruistic responses must be promoted by emphasizing how even large or difficult problems *can* be solved, through small individual acts, in a concrete way.

8

COMPARING THE ALTRUISTIC RESPONSE MODEL TO OTHER THEORIES

I n most cases, the altruistic response model incorporates other theories and findings in psychology and biology into the framework. This model is not designed to conflict with most others, but to integrate and augment them so that we can explain a wider variety of phenomena. Most theories of altruism fit into two camps: ultimate-level versus proximate-level theories. According to this classification scheme, devised by biologist Ernst Mayr, ultimate-level theories explain how a behavior evolved and why it is adaptive, whereas proximate-level theories explain how it is built into the brain and body, the conditions under which it is released, and how it develops within an individual's lifetime. To clarify how the altruistic response model relates to existing theories of altruism, I address the most popular ones in what follows, first with ultimate or evolutionary theories and then with the proximate or neurobiological and psychological theories.

EVOLUTIONARY THEORIES

Almost every article about the evolution of altruism refers to two complimentary, dominant, long-standing theories: (1) inclusive fitness and (2) reciprocal altruism. The altruistic response model is consistent with these views. The tenants of inclusive fitness are indeed critically important for explaining the adaptiveness of an urge to care for offspring and close others. However, inclusive fitness and reciprocal altruism were invented to explain how *any* individual, of *any* species, could benefit from helping someone other than themselves (*including* relatives) even if one's own genes are "selfish." Many extensions of these theories are applied to non-human animals, particularly eusocial species that live in highly interrelated social groups, such as bees, wasps, and naked mole rats. The altruistic response model is different than these because it was designed to explain *human* altruism, based on what we already know about the evolution and inner workings of the mammalian brain. The altruistic response model was particularly designed to explain heroic and active form of human altruism that assist unrelated strangers who are unlikely to return the favor—the very cases that are poorly captured by these two dominant theories, which I address next.

Inclusive Fitness

Inclusive fitness was first described in detail by William Hamilton in his seminal paper in 1964, "The Evolution of Altruistic Behavior."[1] According to "Hamilton's rule," it benefits a giver to help another, as long as the recipient is at least somewhat related to the giver, with the degree of benefit increasing with the degree of relatedness. His formula was simple: "If the gain to a relative of degree r is k-times the loss to the altruist, the criterion for positive selection of the causative gene is k > 1 / r." For example, if you have children, each share half of your genes (the other half contributed by the other biological parent), ensuring that your genes will at least last into the next generation. Moreover, if you assist offspring throughout their long developmental period to reproduction, and they also reproduce, and then at least a quarter of your genes will persist for two generations. If you assist your *grandchildren*, who survive until reproduction, an additional eighth of your genes survive yet another generation. And so on. Thus, genes are not only furthered by the initial act of reproduction, but their longevity also increases with effective caregiving for years or decades after the birth. The specific form of inclusive fitness that refers to assisting one's own relatives was called "kin selection" by John Maynard Smith, who emphasized the role of this process in the promotion of human altruism.[2]

Hamilton did not set out to explain how and why people help perfect strangers, with whom we share relatively fewer genes; he only wanted to point out that, *genetically*, it is not totally *irrational* to help individuals other than yourself, even if genes are "selfish" entities that "want to succeed." For example, if you have a child, caring for him or her comes with opportunity costs to yourself, for example giving up some of your sleep, food, career, earnings, and limiting the number of additional mates or children you might have had. Similarly, if you donate some of your retirement savings to a niece to attend college, that gift at least partially undermines your own future financial security. Hamilton's inclusive fitness explains why people—or any organisms—might relinquish opportunities for themselves to partially related others, despite a genetic goal of propagation.

Even if kin selection is widespread in biology, ironically, it has little influence over how people think about human altruism because most lay individuals would not even use the word "altruism" for aid to a relative. The fact that we consider it self-evident that we should help relatives can itself be taken as evidence that we possess a powerful, intrinsic, implicit motive to care for our

own. Thus, arguments in psychology about whether your motivation to help was focused on how the victim felt or whether you
had warm, tender, selfless and compassionate feelings when you
gave are not particularly relevant to inclusive fitness. Neither are
arguments in economics about benefits to your reputation—unless
those affect your genes, as we will see. It is assumed in biology
that your act should come with some benefit, at least in the long
run, for it to have evolved. Because Hamilton's rule was designed
to explain why it is rational for *any* species to assist individuals
with shared genes, it does not explain the full gamut of activities that we consider "altruism" in humans, particularly if you
insist on defining altruistic activities as those that are purely selflessly motivated and directed toward perfect strangers. Inclusive
fitness also does not explain how such acts are instantiated in
our brains and bodies or whether disparate forms of altruism
evolved or are supported by different mechanisms. Thus, by
design, there is a limit to the scope of Hamilton's rule, which
addresses only part of the puzzle of human altruism.

Despite pointing out this limitation, kin selection is actually
fundamental to the altruistic response model, making my view
complementary with this theory. According to the altruistic
response model, we are motivated to attend to, care for, and protect our own helpless offspring precisely *because* it benefited our
ancestors to do so—a motivation that we extend to non-offspring
usually when it is less problematic. Inclusive fitness explains why
a mechanism for protecting offspring affords a more powerful
and instinctive motivation to help than most other proposed
explanations, including reciprocation, looking good to others,
imagining ourselves in their shoes, or making ourselves feel good
or even better. These latter processes surely contribute to human
aid but played a later and less powerful role in our genetic
selection to feel an urge to respond in specific situations. Only

the altruistic response model explains why people feel *compelled* to act under conditions that resemble the need of a helpless neonate (e.g., a distressed and helpless other in immediate need of aid that the helper can enact) and why rescuers behave so much like our assiduous dams. Only the altruistic response model explains why people in our fMRI experiment, who simply read about fictional charities, gave the most money to charities that assisted babies or children, even compared to adults at much greater risk of death.[3]

Reciprocation and cooperation are the other most popular reasons that people believe altruism evolved. In humans, these latter two theories usually assume that altruism emerged in the context of social group life, which arrived relatively late in the game. We turn to these next.

Altruism in Human Social Groups

Multiple ultimate-level theories of human altruism assume that it emerged in humans as a consequence of group life. (Some allow that the mechanisms are shared a bit with great apes and dolphins, who also live in groups and have relatively large, encephalized brains.) The simplest theory is just an extension of inclusive fitness: you benefit from helping group mates who are somewhat related to you. As an extreme example, if your tribe lived on the same island for six hundred years, you would share genes with one another that were derived from the founding members of the island. Group interrelatedness is particularly fitting for species that live in close social groups, like bees that attack intruders at the hive, ground squirrels that alert each other to an approaching fox, or birds that fly long distances to forage for siblings. Humans are not as interrelated as bees, but even some degree of interrelationship could confer a benefit for cooperation.[4]

Extending this idea still further, "group selection" assumes that cooperation benefits the survival of your group relative to competing groups without cooperative propensities, even at a cost to the individual. If our island tribe is generally cooperative and shares food, and fights together against a common enemy, our collective genes will outcompete those of groups who fail to work together. This is particularly noticeable under intense pressures like those from resource bottlenecks like famine, widespread disease or predation, or attacks from neighboring tribes (which can swiftly wipe out large segments of the gene pool). Group selection has received uneven support over the years because people think Darwin said that selection occurs at the level of the individual, in which case genes cannot propagate that help groups at a cost to the individual. This critique is slightly ironic, given that Darwin himself argued for group selection in 1871. Group selection thrives when the pools of genes between groups are spatially segregated, which is not always realistic, but it has the benefit of not being so tied to inclusive fitness and being able to derive from other shared preferences (e.g., for kindness, similarity, or any phenotype linked to cooperation).[5]

Because these pressures rely on group processes, it is assumed that issues with group interrelationship or group (also called multilevel) selection arrived later and exert a less powerful influence than altruistic responses that originated to protect helpless, related offspring. These ultimate-level explanations also do not afford a proximate explanation for cooperation and do not segregate cooperation by type of aid, overlooking important ultimate and proximate differences. For example, food sharing in chimpanzees and vampire bats is similar in that one individual with plenty lets a weaker individual with less share in their success.

Both phenomena also rely on fairly implicit proximate mechanisms. But only in bats does the sharing require active aid that does not seem to require a large or complex brain; in which case, it is unclear why such cooperation is uncommon in, say, group-living monkeys.[6] In contrast, cooperation during a hunt to corner a prey species seems to require greater coordination, and is usually only reported in great apes, dolphins, and humans. Our human capacity to stockpile and redistribute food to feed the poor or to survive shortages requires more intense effort to ensure equity, avoid theft, detect and deal with defectors, and keep the benefits outweighing the costs. These disparate forms of cooperation appear to emerge in different species and ecologies, at different stages of evolution, through different cognitive requirements—distinctions that are not addressed by theories that assume cooperation writ large emerged as an adaptation to human group life.

By definition, you cannot cooperate unless there are others around to cooperate with. As such, pointing out the relevance of a group context begs the question. Theories that rely on group cooperation also do not address why caregiving and altruistic responses look and operate so similarly and are shared with distant species like rodents—species that also feel peers' distress and provide help. Conversely, some forms of cooperation can be explained through altruistic responding, such as when it was inspired by compassion for the other's distress and need. For example, you might help your neighbor sandbag her property before a storm out of concern for their devastation if the riverbank is breached, or you might shovel your neighbors' icy sidewalk because you would hate to see the frail, older gentleman fall and break his arm. These acts can involve human prospection, ulterior motivates, reciprocity, and good feelings—but they are still motivated by the detection of the other's vulnerability, distress, and immediate need. Because mammals evolved to share one another's emotions, we are trained to reduce their suffering and to bask in their recovery.

Reciprocity is an addition or alternative to inclusive fitness that does not rely upon shared genes within the group. In short: if I give to you, later you may give back to me, making my efforts worthwhile. Suggesting that people suspected a role for reciprocity in altruism for centuries, the Latin phrase *do ut des* (I give so that you might give) appears to capture the same phenomenon. This strategy is extremely powerful in statistical models, even in the most basic tit-for-tat form.

Reciprocity is like group selection in that it works the best in small social groups where everyone is familiar and interdependent (cases where people usually also share genes); however, it is surprisingly less applicable to both our most mundane and our most spectacular forms of human giving. For example, people often donate to charities and help strangers on the street with directions, the time, or to feed the meter without the possibility of being repaid. This is even more true of heroic rescues, which we consider a paradigmatic form of altruism. To address the fact that we help strangers, theorists propose that reciprocation can also arrive *indirectly* from a third party, in a different form, or to other people who share your genes. For example, after donating to the charity or stranger on the street, anyone who observed or learned about your generosity may think well of you, like you more, be attracted to you as a social partner, mate with you, or share with your kin down the line. In hunter-gatherer tribes, a male who brings down a big-game kill during a hunting expedition may generously share his meat with the group; next year, when he comes up short, a family that is doing better will remember him as generous and share with him or his family in return.

Social evolutionary psychologists are particularly focused on better mating opportunities for being seen as helpful—particularly to explain male heroism. This view does not square with the fact that people seem to offer less aid when there are onlookers.[7] People are also the most nurturing in private spaces, and females are the most empathic and altruistic on nearly all surveys and tasks. Displays of altruism are probably more effective in small groups when co-parenting is common and one's behavior is observed (e.g., in a troupe of apes or monkeys or a group of parrots over a solitary, burrow-dwelling vole). Again, as an ultimate-level theory, reciprocity is limited because it does not include its own proximate mechanism, it is not divided by type of aid, and the benefits should arrive later and exert less of an influence on reproductive success than offspring care. Reciprocity also does not explain why we have such similar overt behaviors and mechanisms as our rodent brethren, which exist in different ecologies with much smaller brains. A few people suggest that reciprocity requires a running tally of favors given and received, which would even further constrain reciprocity to humans; this proposal is unlikely given that reciprocation also occurs in other species and can be handled through implicit neuroaffective processes, as we will see.

The "strong reciprocity" model of Ernst Fehr and colleagues assumes that people evolved a nonspecific tendency to cooperate and punish those who do not, because this increased the success of human groups over less cooperative ones.[8] Strong reciprocity is supported by the "Dictator Game," in which half of the students are proposers and the other half are recipients. The two groups do not meet. The proposer is endowed with ten dollars and can give any amount to the recipient, including none, pocketing the rest. On average, people give half of the money to the stranger, representing a sensible point of fairness or equity that seems irrational to economists, given that there was no pressure to share at all (no one should know your decision, the recipient is unrelated, and there's no reciprocation). In the Ultimatum Game, the recipient can accept or reject the proposer's offer, and, if it is rejected, no one gets the money. This encourages fair treatment and demonstrates people's willingness to sacrifice their own rewards in order to punish an uncooperative partner, which is a key element of the strong reciprocity theory. Through functional magnetic resonance imagery (fMRI) and transcranial magnetic stimulation (TMS), the researchers demonstrate how the right dorsolateral prefrontal cortex participates in this process to help people resist personal monetary gain in the service of longer-term fairness.

Strong reciprocity has the benefit over other ultimate models of specifying a proximate mechanism and a context in which it operates. Of course, trust in the games is overstated, since many subjects may not trust the experimenter to be blind to or not to share their decision with others, and there are many other ways that our choice can be discovered (including from us), which invites reputational concerns. Your conscience can also inform you that being fair is "the right thing to do" and feels good. But these methodological concerns beg the question as to *why* we believe it's good to give and bad to be selfish. We do seem to have a strong predisposition toward fairness, but does this *require* a genetic predisposition for cooperation and punishment?

Philip Zimbardo once stated that we learn to obey authorities from our "momma and poppa, the homeroom teacher, the police, the priests, the politicians, the Ann Landers and Joyce Brothers, and all of the other 'real' people of the world who set the rules and the consequences for breaking them."[9] Similarly, early theorists like Sigmund Freud, Hans Eysenck, B. F. Skinner, and Albert Bandura argued that people learn right from wrong from early developmental experiences, such as being praised for sharing and condemned for being aggressive or selfish. These associations between our acts and their consequences are so deeply internalized during development that they are easily evoked later in life, even when we fail to recall our training. Our "internal voice," conscience (as in Eysenck), or ego ideal (as in Freud) learns through affectively laden processes to make the "right" decision in morally loaded situations—like when the cartoon angel and devil sit on opposing shoulders. When we ponder what to do, our bodies quickly and often implicitly produce salient affective and emotional responses that guide us toward choices that were historically good. We feel proud when we imagine following our inner "angel" and guilty when we imagine doing something naughty that upsets others. As such, cooperation could arise any time you live in a long-term group, which faces problems that require cooperation, with a brain that can learn and anticipate contingencies. Mammalian brains generally have this capacity to learn and to dissociate the rewards of giving from the punishment of suffering. Our additional human capacity for prospection may allow us to play out multiple possible options and consequences in advance, or to inhibit an instinctual response for longer-term benefit. This prospection, however, still sits atop a more basic and powerful mechanism that is shared with other species and that does not seem tied to human groups or cognitive skills. Even young children quickly learn to tailor how they count and hide during hide-and-seek so that the game is just challenging enough to be fun and to avoid angering players who feel cheated. This tightrope balance requires a brain that can learn and prospect as much or more than it requires a particularly cooperative brain.

Taken together, strong reciprocity is supported by behavior and includes a proposed proximate mechanism and context.

However, because this theory is tied to *human* social group life and often refers to advanced cognitive skills, I assume that this pressure arrived later and exerts a less powerful influence on behavior than offspring care. It is possible that strong reciprocity per se is not genetically encoded at all.[10] Thus, cooperation can reflect an extension of the basic underlying caregiving mechanism, which arrived later.

Summary of Evolutionary Theories

Inclusive fitness and reciprocal altruism are the two major theories proposed to explain altruism. Each contains sub-versions that increase the scope of the proposal. There is significant evidence to support both theories, across species, and the two do not conflict with one another or with the altruistic response model. Inclusive fitness is critically important to the altruistic response model because it explains *why* it is adaptive for parents to provide responsive, sensitive, and extensive care for offspring, who share half of their genes. Reciprocal altruism can involve the offspring-care mechanism any time we help or cooperate out of a sense of compassion rather than from an explicit strategy or accounting. Regardless, the altruistic response model is assumed to have provided an earlier pressure that was more pronounced, had a larger impact on the genome, and exists in a wider variety of species and cases, particularly compared to forms of altruism that are thought to have evolved in human social groups and require extensive cognitive processes. These ultimate explanations have the benefit of being broader and, so, can apply to any species and situation; however, this breadth is also their weakness. By being so general, they cannot explain how *specific* forms of human altruism evolved or are instantiated in the brain.

Many species help and cooperate, in many different ways. Inclusive fitness and reciprocal altruism are fantastic general tools for explaining aid across contexts. A more precise tool is needed, however, to explain human altruistic responding per se— particularly the heroic type that is directed at total strangers who do not share genes and cannot reciprocate in the absence of extensive deliberation.

PROXIMATE THEORIES

One limitation that arose repeatedly with ultimate level theories of altruism was that they did not explain how the aid is built into the body or brain, when it is set off, and how it differs by type of altruism. We should not assume that the proximate mechanism for activating and promoting aid in humans is the same as that for amoebas creating stalks, ground squirrels making alarm calls, or vampire bats sharing food. We should not assume that the human urge to rush to a drowning stranger is similar to how chimpanzees allow subordinates to grab a stick from their pile. To establish the utility of the altruistic response model, the model must be consistent with data from both existing ultimate and proximate level theories, particularly when describing the same phenomenon or type of altruism.

The altruistic response model is particularly consistent with the social psychological work on how authority figures or bystanders limit altruism, in that both activate an avoidance response that inherently inhibits an approach. This book has not discussed empathy as much, but the altruistic response model is designed to be consistent with theories of empathy that are described in social psychology. However, I emphasize that empathy is not always subjectively experienced in cases where an observer truly feels compelled to rush to help in an emergency, and is more prominent in situations that unfold over space and time. The altruistic response model is tightly linked to existing theories in social neuroscience on how the brain supports giving through reward-based decision processes; however, I focus more on the role of motor preparation than others.

There have been multiple proximate-level explanations in psychology and neuroscience for what causes people to help strangers. There was an early focus on the role for authority figures and bystanders to explain our marked apathy in historical cases of cruelty and suffering like mass genocides. Later, researchers tried to explain when we can be empathic and truly focus on others' need. Recently, researchers have begun to integrate research on empathy and altruism with more domain-general research on decision making, to show how learning, probability, and emotion support decisions to help. The altruistic response model is consistent with these existing proximate explanations, but because it was designed to integrate ultimate and proximate views through offspring care, it diverges from prior work in a few ways. For example, only the altruistic response model assumes that distress and stress can promote aid when the observer knows what to do and it does not require conscious, deliberated decisions to act. The altruistic response model also makes specific predictions about who will help, based on the situation. As caregiving mammals, in circumstances that mimic our need to protect helpless offspring, people can naturally proceed from observing to helping victims, without much thought between the two; however, when there is plenty of time to decide, this intuition can merge with feelings of empathy and thoughtful consideration that people traditionally assume direct our aid.

Susceptibility to Authority

The largest and most relevant body of work to altruistic responding was ironically designed to show how unhelpful people are toward strangers in need. This tradition emerged from a post–World War II movement in social psychology aimed at discerning how people could have participated in and observed the mass genocide of Jews before decisive action was taken to stop it. These researchers were less concerned about nefarious dictators striving for world domination and more about how so many regular citizens stood by and watched—or even participated—despite construing the acts as wrong. As caring humans, we like to think that if someone in our environment were hurting and killing innocent people, we would say something. We would like to think that if someone asked us to help them commit mass homicidal atrocities, we would refuse. However, research (and real life) has demonstrated that our better angels do fall under particular conditions.

Stanley Milgram demonstrated that even typical American adults, from good homes, without any financial incentive or risk to their own safety, would follow the instructions of an "authority figure" (really just a teacher).[11] All forty subjects in the first study agreed to the authority's instructions to administer electrical shocks to a confederate who missed memory questions at the initial, lowest level ("slight," 15 volts); they also continued to agree as the voltage increased through "intense shock" at 300 volts. Some percentage of students did refuse to obey with increasing shocks, but almost 70 percent still administered the highest-level shocks—"danger-severe shock" or 450 volts—despite the fact that there was no punishment for disobeying and the confederate was clearly suffering. My title, *The Altruistic Urge*, seems to fly in the face of such clear evidence that we are apathetic, even cruel. Milgram's data, though, are perfectly consistent with the model through the fundamental tension between approaching and avoiding the victim. People do not naturally help when they are inhibited or scared. Moreover, Milgram's students were more likely to agree to the shocks when the victim was far from them and they went to great lengths to avoid feeling responsible for or experiencing the victim's pain, for example, by looking away or touching the dial as lightly as possible—random details like those provided by Wilsoncroft that would not be allowed in a modern report, even though they actually attest to our sensitivity.

Zimbardo's Stanford prison experiment is usually mentioned in the same breath as the Milgram studies, in which similarly well-educated students at Stanford University became participants in a mock prison where they were randomly assigned to be prison guards or prisoners.[12] The original story reports that guards spontaneously engaged in bullying and dehumanizing their peers, who had sat next to them in class just days before (even more so at night when they were less observed).[13] Like Milgram, Zimbardo presented these results to prove that even "good" people, who were socialized to value compassion and treat each other with respect, possessed the capacity to be inhumane when placed into a position of authority. Both researchers also emphasized that people license such cruelty by reframing themselves as not being responsible for the harm, since they are only following orders—shifting the cause of suffering onto the authority figure and exonerating their guilt.

These studies and their attendant theories are consistent with the altruistic response model because the model states that people do not feel the urge to help when frightened or otherwise incentivized toward different or competing goals. In the real world, people who are called into action by authority figures are often threatened with death to themselves or their families if they do not follow instructions, and the soldiers often rely upon the authority for money, protection, or food during periods of great unrest. Humans also exist in a hierarchy like other social mammals, which is maintained through a long series of rewards and punishments. For example, Frans de Waal wrote extensively about the strategic power plays in chimpanzee social groups, including how dominant males can apply (sometimes deadly) force to secure power.[14] Strict hierarchies also abound in human society, including on school playgrounds.[15] Thus, the harm that people bring to others is not so unpredictable if you assume that we too are social, learning animals who track rewards and punishments over time, to make choices that improve our own fitness. Soldiers need not be *enthusiastic* about carrying out instructions, though surely some are: 5 percent of Milgram's subjects chose severe pain even when they were allowed to select the pain level. The dehumanizing way in which people often institute their cruelty is itself support for our natural tendency to care about others' suffering, because by first making them seem less human, we do not have to share in and regret their pain.

Bystander Apathy

In this same post–World War II era of social psychology, John Darley and Bibb Latané examined why people stand idly by while others suffer.[16] In a typical bystander apathy experiment, a confederate research assistant might feign injury or pain in a public place, or a laboratory participant may hear someone's pain or distress from an adjacent room while performing an unrelated task. The experimenters measure how long it takes participants to approach the apparent victim and offer help. Typically, participants respond less to the degree that there are bystanders present, which is called "bystander apathy" or the "bystander effect" and is thought to reflect a "diffusion of responsibility" in groups. In a group of observers, no one person is responsible for helping because there are many other possible helpers, some of which may be even more qualified. This phe-nomenon is often perpetuated in television shows like *Dateline* or *What Would You Do?* in which people are filmed in public places passing by an actor feigning injury or who is passed out to demonstrate our widespread apathy. This research is often placed into context with real-life events like the murder of Kitty Genovese in New York. Early newspaper reports stated that thirty-eight neighbors heard her screams or witnessed the act and did not try to help, as a former boyfriend stabbed Kitty forty-three times in the alley behind her building.[17] Kitty's story was offered as an example of modern life, wherein we are surrounded by strangers and feel no connection to or compas-sion for them.

There are multiple points of contact between Kitty's case, bystander effects in experiments, and the altruistic response model. It is important to look carefully at the data, however, and not just accept the implications in news headlines or article titles. For example, in Darley and Latané's studies, people did actually help when the situation was more natural, such as when they were free to react, were not instructed to stay in the room, or met the victim earlier and could connect with them. Nearly every sub-ject in their 1968 study who faced the victim alone helped—the vast majority in fewer than ten seconds—and they reported no particular intervening thoughts between hearing the distress and responding. Even in the Holocaust, people were more likely to hide and assist victims in their homes when they were not targeted by soldiers, observed the victim's distress, and felt empathy, love, compassion, or a responsibility to help. Under these conditions, altruists report feeling compelled to act, without "considering risk or thinking about being either lauded or maligned."[18] In the case of Kitty Genovese, her neighbors were not quite as apathetic as originally described; for example, a dozen or so people may have heard screams, but not clearly or for very long, and one man did call out to help.[19]

People surely experience a "diffusion of responsibility," but their inaction in many of these circumstances is not irrational. For example, intervening with a knife-wielding manic in a dark alley at night can surely bring injury or death . . . at the very least a protracted and dangerous new role as witness to a violent crime. The experimenter may chastise you or refuse course credit if you fail to acquiesce to their study instructions, which are threatening prospects for people living comfortable, upper-middle class college lives. Someone writhing on the street may have a terrible transmissible disease or could be unstable and trying to trap you in a ruse. When you help someone in a medical emergency you could make things worse or be punished, taken advantage of, or even killed. Even calling the police is not without risk, as we have witnessed the tragic consequences of calling officers to uncertain and volatile situations (that might not even be problematic), who then make things worse with their fear, racial bias, and guns. Thus, people may not be as altruistic as we would like them to be, and they may be influenced by authority and inhibited by onlookers, but they quickly and implicitly calculate the risks and rewards of intervention and adaptively avoid rushing in when they are not capable. Through the opponency between approaching and avoiding, which sits at the core of the altruistic response model, we can both explain why people do not help in dangerous situations but also when and why they *do* intervene (e.g., when they have expertise, directly observe the victim's distress, are personally connected to them, and so on).

After the post–World War II surge of studies on altruism (or our lack thereof) in the late 1980s and 1990s, social psychology shifted to examining our "better angels," through testing the relative influence of the person versus the situation, the emotions that underlie altruism versus avoidance, and how both are impacted by temperament or personality. We turn to an overview of that work and how it relates to the altruistic response model next.

The Empathy-Altruism Hypothesis

Much of the new psychology of altruism resulted from the prolific work of C. Daniel Batson, a former theologian turned experimental social psychologist. Batson used a series of clever experiments to demonstrate that people are actually capable of "true altruism"—meaning that they do sometimes help others out of a truly compassionate concern for the other.[20] This line of work aimed to combat the growing (and still common) belief that, deep down, people are selfish and only help when it is in their own best interest. Disregarding the false dichotomy between being selfless and benefiting from an act, Batson described how people will help when they feel a sense of "empathic concern" (sympathy, compassion, tenderheartedness, warmth) for the welfare of someone they value and focus on their need. In contrast to the promoting effects of empathic concern, people were thought to help only when they had to if they felt "personal distress" (troubled, worried, distressed, etc.). Batson's dogged work to demonstrate this link to other-oriented helping bred a cottage industry of experimental research on the motivational basis of human empathy-based altruism, which continues today.

A parallel line of research in developmental psychology came to similar conclusions for children. Nancy Eisenberg and Carolyn Zahn-Waxler and colleagues found that children help when they feel other-oriented sympathy for the other's needs, which develops relatively early in a child's life.[21] Even toddlers show attention and concern for another in distress or pain and often take actions to try to alleviate their distress. It was particularly enchanting that Zahn-Waxler reported that when a caregiver feigned distress in the home, the family dog would often respond, approaching and attending to the caregiver, whining, or placing its head on the parent's lap.[22] This brief comment provided an early scientific record of nonhuman concern and help, which has since been studied in more detail with experiments in multiple species, including dogs, apes, and rodents.[23]

More recent research in this tradition has demonstrated the degree that we share in others' pain at the neural level. For example, India Morrison and Tania Singer have separately demonstrated that the anterior cingulate cortex and anterior insula are activated in the brain when people directly experience physical pain and observe another's pain.[24] This effect has been replicated dozens of times.[25] Extensions of this work have shown that people experience less shared pain for people in their out-group (e.g., culture, race, or even soccer team) or after habituating to the pain over time (e.g., in surgeons).[26] We demonstrated at the University of Iowa—with Antoine Bechara, Antonio and Hanna Damasio, Tom Grabowski, Brent Stansfield, and Sonya Mehta—the same neural signature and emotional arousal when you imagine your own and another's emotional experience of anger or fear, unless you cannot relate to their experience on the basis of your own.[27]

Generally, this research is consistent with the altruistic response model, because of the agreement that an empathic state of concern motivates aid and evolved to foster aid toward off-spring and was later extended to other group members and occasionally strangers. These frameworks also commonly assume that you feel more for and help those who are similar, bonded, familiar, like you, in your group, and when you know what to do. Researchers also generally agree that people process others' pain or distress through their own neural substrates for feeling this way, even if they disagree about what happens next or how consciously we are aware of this shared feeling. (I do not think there is always awareness.)[28]

Despite broad agreement about the utility of shared affect and empathy for altruism, the empathy-altruism theory appears to conflict with the altruistic response model. Emergencies that make people feel stressed, aroused, or distressed from observing the victim's distress should not promote aid according to the empathy-altruism hypothesis, but they can in my model. This paradox reflects different meanings of the word *distress*. Batson's "personal distress" refers to a subjective, upsetting emotion or motivational state that is almost always measured when you cannot directly help the other, who does not require immediate aid. In contrast, distress in the altruistic response model refers to the obviousness of the others' genuine and immediate need, which signals their urgent and aversive problem, that you need not share in consciously. Your brain only needs to correctly process their state as distress, through similar neural representations refined through your own personal experience; this does not mean that people usually become distressed or panicked in a way that prevents action, as implied by others.[29] Moreover, people with expertise can snap into action rather than fall to pieces in emergencies because they are prepared—a key tenet of the altruistic response model. Thus, empathy-based views of altruism and the altruistic response model are convergent if you attend to the mechanism, but only the latter can explain why people enact heroic rescues quickly, without feeling empathy, in a way that is facilitated by activation of the sympathetic nervous system and stress hormones.[30] The altruistic response model was therefore designed to explain cases that are hard to explain through other models, such as why people sometimes *do* help, even in dangerous and heroic situations, without feeling calm or compassionate, toward a victim who is not related and cannot reciprocate.

Affect, Decision Making, and Neuroscience

Part of the altruistic response model involves decision making. In psychology, usually different people study empathy or altruism versus decision making, but sometimes the two converge in behavioral economics and social neuroscience. Theories in social affective or decision neuroscience do not refer to our learned morality or superego as older theories did, but still assume that people are implicitly (and sometimes explicitly) guided by neural predictions of advantageous choice, signaled by affect, that is trained through past experience.[31] For example, when you play the Dictator Game and try to decide how much of your $10 to share, you may feel a pit in your stomach imagining how mad your peer will be after learning that you gave nothing (or how your friends will admonish your selfishness if you tell them about it) and in turn decide on an even split. If you want to feel like a great, giving person, you might focus on how you both will bask in the great glow of gratitude after you donate all of the money. If you are hungry, you might focus on the delicious sandwich you could buy afterward. If you are poor, you might consider the water bill that needs paying. There are many inputs to your decision, which are weighted differently by each individual, that integrate past and current conditions.

Regions of the brain in the mesolimbocortical system that also participate in offspring care are activated in these experiments. For example, the OFC, NAcc, insula, and amygdala are typically activated more when people make altruistic decisions in the scanner, with the OFC more often being involved during conscious, deliberated decisions and the NAcc in response to rewards. Some researchers even interpret NAcc activation when people punish a defector in an economic game as a sign that retribution "feels good" (though this interpretation isn't particularly well supported). These theories and data are consistent with and inform the altruistic response model, but the current model is more focused on altruistic responding per se. As such, the altruistic response model assumes that the urge to respond can occur implicitly, without the conscious deliberation usually suggested for human decision making, and it makes more detailed predictions (e.g., that the MPOA, sgACC, and neurohormones support the urge to help in delineated conditions). The current hypothesis is also distinguishable from existing neuroscientific views because it is more integrated with the ultimate-level descriptions described earlier for how this urge evolved and why it is adaptive.

SUMMARY

There are a few major existing theories for how altruism evolved and is instantiated in our brains and bodies. Generally, the altruistic response model does not conflict with these theories—it even relies heavily on inclusive fitness or kin selection and views of how affect informs decisions in the brain. Most of these theories are either ultimate or proximate explanations of behavior, without considering how the proximate mechanism itself evolved and is adaptive across species, and without specifying conditions under which aid is released. Most ultimate theories are domain- and species-general and focus on explaining eusocial or highly related nonhuman animals or how humans evolved to cooperate in groups. Most proximate-level theories focus on conscious deliberation and find evidence for such contemplation, but only because it is baked into the design of the studies, further perpetuating this focus. The altruistic response model is the only one to address forms of giving outside of the central tenets of most theories of altruism, that is, that we should be related to the victim, reciprocated for our act, and consider acting like the thoughtful humans that we (hope we) are. The altruistic response model assumes that decisions to help need not be conscious, and rely upon machinery that is available across species—especially when there is an urge to help that resembles responding to our own helpless offspring.

CONCLUSION

Why Consider Altruistic Responding Now?

This book has explained the altruistic response model, in which human altruism—even the heroic kind toward complete strangers—can reflect our ancestry as caregiving mammals who had to respond quickly to slowly developing, helpless offspring. This is why people feel an urge to help victims who resemble this context, such as those who are young, helpless, vulnerable, distressed, and need immediate aid that is feasible. This process is subserved by ancient neurohormonal circuits that ensure a fast, intuitive response under such conditions, which were historically adaptive and do not require human-specific cognitive capacities.

In maternal rodents, this active response is triggered by the process of becoming pregnant and delivering pups, accompanied by a host of neural and hormonal changes that transform dams from avoiding pups to expressing a potent, driving motivation to approach and care for them.[1] This does not mean that only females who have given birth help. Even in laboratory rodents, males and nonmaternal rodents can also provide care, as long as they have had time to adapt to the presence of the novel pup.[2] Both dams and nonmaternal rodents provide care through a shift in the underlying neural circuitry between avoiding and approaching.

Even if males and young females can provide care once properly prepared, this ancestry in maternal care also explains strong gender differences in human altruism. Most passive, succoring, and tireless forms of human giving are biased toward females, whereas our most publicly lauded heroes are more often males who have the right combination of strength, expertise, and risk-seeking to intervene.[3] Many cultures teach females to be caring and men to be brave—which may explain some of these gender differences—but these differences also likely reflect our retained, neurobiological origins as caregivers and as protectors. As evidence, even in cultures with relatively high gender parity, in heterosexual, two-parent families where both parents work, the mothers still provide significantly more offspring care than the fathers.[4]

As was the case with William E. Wilsoncroft's dams, introduced at the beginning of this book, I argue that people also possess a natural opposition between avoiding and approaching victims in need. People avoid approaching when they feel overwhelmed, scared, incapable of helping, or uncertain about the victim's motives. But they switch to approaching in situations that resemble offspring care, with a vulnerable, helpless, distressed victim who needs immediate aid that the observer thinks he or she can successfully provide. Observers are even more likely to approach when the individuals are bonded to each other like the dam to her pups. This neural and psychological opponency ensures that we only respond when our generous spirit is unlikely to be taken advantage of, given that we are most comfortable around bonded social partners and rarely encounter truly helpless adults.

Our perceptual and cognitive systems are also designed to make accurate predictions about whether need is truly urgent, what the proper response is, and whether we can enact it in time. People rarely leap into icy waters when they can't swim, run into burning buildings when they're too weak to carry someone out, rush toward people with medical problems that exceed their knowledge, or even engage tearful friends whom they might only further agitate. Our avoidance in these cases ensures that we care for our own related offspring and family—and sometimes strangers—without unduly undermining our own survival or fitness. At the same time, the implicit neurophysiology of this shared avoid-approach dichotomy means that we can make these predictions without any conscious calculations. This further explains why heroes report "just responding," without stopping to contemplate the costs and benefits, whether they will be repaid, or feelings of sympathy and compassion—factors that people often assume promote empathy and altruism.

Even though I describe the aid as emerging from an "urge," that does not render the response stupid, simple, or reckless. By design, the response occurs only when it is adaptive, on the basis of quickly integrated information about the victim, observer, and situation. Even when "instincts" are described by biologists for species we already assume act instinctually—amoebas, fish, birds, rats—the responses are still only predispositions, embedded in epigenetic systems, that require multiple genes acting in tandem, which are affected by early development and the current context. Thus, even a rather base or instinctual act still includes complexity that is sensitive to context and allows for individual variation. Moreover, because instincts must be encoded through a specific biobehavioral mechanism, if we understand the mechanism, we can predict both normal and expected responses alongside strange and unwanted ones, such as when geese retrieve volleyballs instead of their own eggs or people die in an icy pond rescuing their neighbor's dog.

As explanatory as the altruistic response model is, it applies only to situations that release this active, intuitive urge to respond. The model does not explain—does not attempt to explain—all forms of altruism. Altruism is a broad category of behavior, from worker bees that digest their queen's food to EU nations that coordinate to help oppressed people thousands of miles away. Theories of altruism that claim to explain all of human giving tend to categorize all costly or giving acts as altruism and define forms of altruism by how they look from the outside. The altruistic response model carves nature at its joints and instead segregates forms of altruism by how they evolved and how they are processed in the brain and body.

People might feel emotionally or psychologically attached to the "specialness" of certain types of altruism, which seem specific to humans or involve sympathy, compassion, empathic concern, or an ability to ponder how the other feels. Such forms of aid clearly exist, as decades of research on empathy-based altruism or perspective taking in humans can attest. But even an urge that is not specific to humans can still resonate with our need for altruism to entail warm, tender, positive feelings, like the ones we experience in close contact with bonded friends and family. Thus, the altruistic urge is still "warm and fuzzy"—it's just that the rewards are not monetary, and you need not consciously bask in them before deciding to act. Good feelings are a normal, adaptive consequence of a rescue that reinforces the desire to repeat the act in the future. Such a reward need not detract from construing the act as "truly altruistic," since the reward is part and parcel of a mechanism that persisted in the genome specifically *because* it benefits the giver.

WHY NOW?

There are dozens of books and thousands of articles about our human capacity to help. I could have written about lots of different topics somewhere in this vast space, from my own view of empathy to an overview of theories of altruism to a tome on human goodness writ large. Why write about this? Why now?

This specific homology with offspring retrieval was important to describe because, for one, the descriptions of offspring retrieval and its neural bases so resemble our own altruism, particularly for heroic forms that were previously not really explained. The homology was also important because the intrinsic opposition between avoiding and approaching helps us understand why we are both apathetic bystanders and empathetic helpers and why people describe humanity as either grossly selfish or amazingly generous. Through the natural dichotomy between avoiding and approaching others—derived from dozens if not hundreds of studies on caregiving in multiple species—we can merge our assumption that we evolved as caregivers to be empathic and responsive *along with* a failsafe mechanism that forces us to stand idly by when it is dangerous or we might not succeed. If we do help an unrelated stranger, even through an urge, we should not consider it a "mistake" or "error," because the act was issued through implicit processes in our bodies and brains that evolved over millions of years to help us survive.

Heroism is also the least well studied or understood form of prosocial giving. Heroism is very different than the types of altruism that people usually study, like alarm calls, social grooming, consolation, food sharing, or giving a few dollars to a stranger in the lab. It is hard to replicate heroism in a controlled experiment, particularly during brain imaging. The altruistic response model allows us to understand how putting your arm around a crying friend and leaping into a river to save a stranger can both emerge from the fact that it was adaptive to want to protect our own.

This altruistic response model is consistent with existing theories of altruism like inclusive fitness, reciprocity, sexual signaling, bystander apathy, and empathy-based altruism. However, because this model is based on animal models from neurobiology, it can be more specific, more so even than other theories grounded in caregiving. According to the altruistic response model, features such as felt compassion, group coordination, cost-benefit analyses, or simple grandstanding exist and benefit people, but they were not the *primary* or *initial* motivation to respond, because mammals cared for dependent offspring long before these other features were possible or useful. The altruistic response model explains why such similar acts of aid are observed in species with much smaller brains such as rodents, and even birds and ants.

For these reasons, I believe that we have the most to gain right now from applying a deep wellspring of information on the neurobiology of mammalian offspring care and retrieval to human altruism. This altruistic response model aims to fill gaps in our understanding of human altruism while merging existing theories into a larger, more coherent framework that can explain some of our strangest acts of love.

Some of the ideas in this book may turn out to be wrong. The theory would fall onto shaky ground indeed if it turned out that there was something catastrophically wrong with the animal models of offspring care. Thankfully, this is unlikely, given the number of different researchers, methods, and species that informed the model. The altruistic response model is not subject to single bad-apple researchers or statistical techniques that require psychologists to retract so many of their "effects" of late. Even if some parts of this model need to be revised in the future, the fact that I provided specific, testable hypotheses allows us to advance our understanding beyond where it too often remains: in shrouded speculation.

THE FUTURE OF ALTRUISM

There are many aspects of altruism that I feel fairly confident about. I feel comfortable that I can predict what the average person will do, and which brain areas will be involved, in a wide range of situations that people often face. I understand which stimuli powerfully direct our behavior and when people will feel the urge to act versus the predilection to hold back. I understand how more motivated forms of altruism differ from more reflective ones. Yet, there are a few aspects of the model that have not been demonstrated, particularly in human aid to strangers, and there are aspects of human goodness and morality that I have left unresolved intentionally here.

The MPOA as Necessary and Sufficient

Research on humans has yet to link the most necessary neural region in dam pup retrieval—the medial preoptic area (MPOA)—to altruism. Demonstrating a link to this specific area is difficult because the MPOA is a very small nucleus, deep in the middle of the brain, which is hard to access through scalp measures like electroencephalograms (EEG) or transcranial magnetic stimulation (TMS) and hard to locate with functional magnetic resonance imaging (fMRI). Moreover, if the MPOA is truly specific to retrieval, then it would be activated only in situations that very closely resemble a physical retrieval or that involve the safety of kin, which is hard to approximate in a brain scanner where people must lie completely still while presented with many repeated, short events.

Species Differences

Even if mammalian brains are clearly homologous, they also diverge when it comes to very specific details like the precise number or distribution of neurotransmitters, neuromodulators, hormones, or receptors in each brain area in the circuit. For example, in the animal model of bonding between mated females and males, there are more than six times more oxytocin receptors in the nucleus accumbens of monogamous prairie voles than in nonmonogamous montane voles.[5] Thus, even if regions like the accumbens and the frontal cortex similarly work together to motivate aid in rodents, primates, and humans, we expect species differences in these interconnections based on each species' ecology and mating system.[6] As was true of the small but essential part of the hypothalamus for offspring retrieval (the MPOA), these species differences will take time to document, since we are severely limited in our measurement of such small brain areas in alive, awake humans.

Neural Correlates of Parochial Altruism

Research has shown that people are more sensitive to the needs of familiar and interdependent victims. This has even been demonstrated in rodents, which will free a familiar trapped rat or help a rodent of their own species (or the species they grew up with) over unfamiliar rodents.[7] People are also strongly biased toward in-group members, defined in multiple ways, including race, ethnicity, nationality, or which university or even soccer team they support. There is literally less empathic pain in people's brains when they observe the pain of an out-group member.[8] People of color are systematically given less pain relief in the hospital.[9] We are also less swayed by others' pain when we aim to hurt them to satisfy our own needs, compared to when we passively observe our partner's pain. Sometimes disengagement is beneficial, like when a doctor habituates to or reframes a child's pain from an inoculation that prevents disease or when a parent ignores a child's cry so that both can get some sleep. Researchers have studied the end result of our parochial altruism, but not how we got there.

The perception-action model of empathy explains why we are more empathic toward those who are similar and familiar, by relying on the fact that our representations of other people, affective states, and situations are refined through experience and that we benefit as social, caregiving animals from helping familiar, interdependent in-group members who can return the favor and whose fate is tied to our own.[10] We pay more attention to these people and better understand and imagine how they feel. In addition, simple morphological features like skin tone, age, style of dress, or gender are easier to process when we have more experience with them—when we grow up perceiving and processing them. Moreover, people can ignore others' pain or need when it is irrelevant to their goals, all the more so when they want to compete with or harm the other. Executing your own goal-oriented action and carefully observing another's state conflict at a neural and psychological level, such that you can't do both effectively at the same time. Thus, we know *some-thing* about how we come to favor those who resemble ourselves, but the precise neural cascade that subserves this response is poorly understood—we have only documented the regrettable outcome.

To subvert this "hardwired" tendency to feel for those like us, and not for those unlike us, we need to provide people with rich, positive experiences with out-group members, avoid stereotyped and misleading media, and stop relying on biased, subjective judgments for things like felt pain. For example, medical profes-sionals should follow rule-based decision matrices to determine the level of pain of the patient, without emphasizing their sub-jective and biased view.

Links with Higher-Level Reasoning and Morality

Most alternative theories of altruism assume that the need to collaborate with group members and consider someone else's thoughts (regarding what you are thinking and so forth) drove the selection for larger brains, moral intelligence, and intelli-gence writ large. As one of my favorite songs by my husband goes, "I wish I were so sure." The altruistic response model only aims to address forms of altruism that do not rely on higher-level cognitive processes, because it assumes that cognitive capacities emerged later and sit atop the preexisting capacity for aid. Through this focus, I have intentionally not specified when and how I *do* think these later cognitive capacities evolved, as I have yet to be convinced by any one argument about the specific selection pressures that created large, thinking brains.

One reason that I am suspicious of these models is because they are primate-centric and assume big brains are needed to be cooperative, altruistic, or smart. However, even birds, which possess literal "bird brains," are capable of many of the same feats, more of which are discovered each month. Thus, perhaps social intelligence is more tied to social ecology than to the primate taxon or large brains. Trivers argued that cooperation could emerge spontaneously and be a stable strategy under the right conditions; for example, if you came across the same individuals repeatedly and had a long developmental period that involved caregiving and neural plasticity; however, he did not specify how this could occur at the level of the brain and body and whether the machinery would differ between mammals versus birds.[11]

For similar reasons, I stated that the altruistic response model applies to caregiving *mammals*—because I am more confident in the homology within mammals. But altruism exists in other species, including insects and birds. Birds are capable of social bonding, deception, food sharing, tool use, episodic memory, contagious stress, and mirror self-recognition—all processes that people assume require human "consciousness." African gray parrots not only can imitate human speech but also can learn words that refer to objects and their properties similar to the way that human children do, including objects that are not present.[12] If you need a large brain, with a high degree of encephalization, to perform complex social and practical tasks . . . then how do birds do it? The contagious hormonal and autonomic stress that we have demonstrated in humans has also been found in pair-bonded zebra finches and even baby chickens.[13] Even ants free other trapped ants through a mechanism that is highly similar to our own contagious stress mechanism, with the trapped ant releasing a stresslike hormone that elicits stress in the passing (and interrelated) ant, which then helps.[14]

Social bonding and offspring protection seem like better predictors of altruistic responses per se, compared to brain size.

Bird mothers provide extensive care for chicks after their eggs hatch, feeding and protecting them until they are strong enough to fledge the nest. In some bird species, females and males pair-bond and care for the offspring together. Many birds also live in large social groups that may require sophisticated information about one another and their behavior. Alternatively, bird brains may compensate for their smaller size (which aids in flight) with more neurons, which are more densely packed into a smaller space.[15] Thus, maybe you do not need a large brain to be altruistic, but you need a lot of neurons to exhibit cognitive complexity.

The mesolimbocortical system, which was central to the offspring care system, may be conserved in birds even though their brains look different (smaller, rounder, smoother, with somewhat of a different organization). Many neural regions in birds map functionally and structurally onto analogous or homologous regions in humans and rodents, including the avian hippocampal formation for spatial memory and navigation, the striatum (including Area X, a homologue to the basal ganglia) for motor learning and execution—even a neocortex.[16] Birds also have dopamine, which supports bird song through connections with the striatum, pallidum, and motor system, as it does in human learning (see figure con.1).[17] And they possess neuropeptides like mesotocin, which is analogous to our oxytocin. This oxytocin homologue has even been identified in male lizards and is functionally linked in at least one lizard species to mating, like the mechanism for pair bonding first discovered in monogamous voles.[18] In fact, neurosecretory nonapeptides like oxytocin are thought to have evolved more than 600 million years ago, *before* the division of protostomes and deuterostomes, with evidence across species that these molecules subserve the

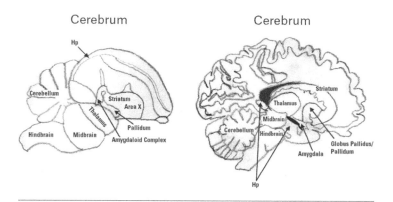

Cerebrum   Cerebrum

FIGURE CON.1 Drawing highlighting striatal, pallidal, and cortical
regions that participate in altruistic responses, which are considered
homologous because they exist and support similar functions
in both humans and birds.

Drawing by Stephanie D. Preston, CC-BY-SA-4.0, based on information provided in
Kristina Simonyan, Barry Horwitz, and Erich D. Jarvis, "Dopamine Regulation of
Human Speech and Bird Song: A Critical Review," *Brain and Language* 122, no. 3
(September 1, 2012): 142–50, https://doi.org/10.1016/j.bandl.2011.12.009.

balance of fluids and the birth process in an egg-laying bird, a
live-bearing rodent, and humans.[19] Thus, even birds, which possess tiny brains and split long ago into a group of sauropsids
from our own placental mammals (synapsids), may possess
homologous neural structures, neurohormones, and neural
functions to those observed in mammals. Given the similarities
between mammalian and bird brains (and to some extent even
reptiles and insects), it is conceivable that aspects of the mechanism for altruistic responding extend beyond mammals. I await
further evidence.

People's assumptions about the evolution of caregiving across
species have been biased by our focus on primates. We assume

from primates (most of which are *not* biparental) that the ancestral form of offspring care was to not provide any care, followed
by female-only care, followed in limited conditions to biparental care (in 9 percent of mammals). In reality, the serial order of
this pattern varies widely in the rest of the animal kingdom.[20]
For example, teleost fishes are assumed to progress from no
caregiving to sole *male* care, biparental care, and then only later
female-only care (female-only occurred in only one species
measured).[21] In cichlid fish and birds, biparental care is usually
assumed to be the *ancestral* state, which sometimes gave rise to
female-only care, whereas in mammals this sequence is reversed.
Even if most birds are biparental, with a few providing female-
only care, some data indicate that male-only care was the ancestral state. Some species even appear to have evolved backward
from the presumed linear sequence, such as in the few cases
where a species evolved from egg laying to live births, only to
later to revert back to egg laying, or in the many cases where a
species became biparental before reverting to sole care by either
the male or female.

Returning to the way in which we understand the homology
of the brain, it seems more realistic to assume that there are
genes, regions, and hormones that are shared across diverse
species—taxa even—that can be altered in their expression,
timing, and proportion (especially during early development),
to yield a wide variety of behaviors. Thus, the proposal that we
share some genes and blueprints for building a brain and body
that responds to offspring need does not mean that any caregiving ancestor will automatically give rise only to newer species that also care for offspring in that same way. Rather, the
brain and body response to mating, pregnancy, giving birth,
and caring for offspring can reemerge, as needed, to suit the
context of each species or individual and can amplify processes

that benefit survival in each ecology, as part of the process of speciation itself.[22] The parsimony lies in the fact that such similar mechanisms across species and taxa are unlikely to have arisen spontaneously and independently each time. Rather, this seemingly spontaneous evolution of caregiving in mammals, which relies on similar brain and body processes across species and taxa, relies upon limited, fundamental building blocks for nervous systems that permit a constrained number and type of variations, through alterations to the timing and sequence of genes in early development. This is like the situation for baking cakes. Almost all cakes require flour, eggs, and sugar, regardless of the occasion or how you want it to taste. To alter the cake to make it appropriate for a wedding or a picnic, to make it rise high or not at all, to make it taste like chocolate or lemon, the baker needs only to modify the proportions of those basic ingredients or add a few additional ingredients that do not change the reliance upon those fundamental, basic ones.

To sort out these remaining issues, we need more collaboration across fields, particularly incorporating knowledge from evolutionary biology, paleontology, genetics, and developmental neurobiology to inform psychological and neural theories. We should ensure that our view of altruism is not yoked to a caricature of evolution that proceeds from primitive, small-brained animals that lack cognition to advanced, large-brained humans that always employ conscious deliberation. We should understand how behaviors that look different across and within species can be produced by relatively small changes in the genome, which nonetheless produce striking variation in attributes that we care a lot about. Based on my expertise, I chose to focus on the potential for a homology among caregiving mammals, but these processes are expected to extend, at least in broad strokes, to other branches of the phylogenetic tree.

FINAL REMARKS

The way that we understand even our most lauded acts must consider that humans are animals. We have shared in a long, evolutionary process that refined the way we perceive, predict, and respond to others, to adaptively solve problems that strongly impact survival like feeding, mating, and raising offspring. Why not apply this vast, accumulated knowledge from other species that, by and large, already exists in the repository of science, to understand when and why people want to act?

The altruistic response model integrates ultimate and proximate explanations of human altruistic responses, across all four of Tinbergen's levels of analysis, which were heretofore largely ignored. This model frames altruism as just another behavior to be examined from an ecological and biological perspective, rather than a special, nice, or uniquely human or cognitive process that is filtered through practical concerns about what we can easily measure in the lab. Most experimental treatments of altruism in economics and psychology involve giving money to strangers in the laboratory—quite unlike the altruism "in the wild" that is experienced by us, our ancestors, and other species. Money even changes people's mindsets when they approach a task.[23] Our mindset certainly differs when we consider giving x or $2x$ of the experimenter's "house money" to a stranger or to ourselves versus when we decide, with no clear alternative, to rush toward a stranger in urgent need. Our early ancestors did not need cost-benefit decisions and were not trading their own for another's reward when they pulled infants, relatives, and interdependent group mates from danger or rushed toward distressed, bonded others in need.

The altruistic response model uniquely emphasizes this overt motor response. Empathy and altruism are often characterized as resulting from high-level, abstract, cognitive feats that require extensive explicit thought and deliberation. We surely do think long and hard about whether to help someone . . . sometimes. However, the brain is designed to learn from experience and to quickly predict outcomes. The motor system in particular is defined by "expertise"—implicitly and naturally producing expert knowledge about what our bodies can and cannot accomplish, the best way to respond, and how quickly it can happen. The motor system is highly predictive, accurate, and intrinsic to decisions to act, in the moment, without needing conscious deliberation. Altruistic responses are behaviors— motor acts that we should understand as such.

Humans are inherently social. We need other people to survive and thrive. We give and we receive. To understand this dynamic, we must not only look at the people to our left and right, but also much further out, across species and into our distant past.

I hope you enjoyed this foray into the nature of human altruism. Maybe the next time you find yourself staring at pictures of cute puppies on the internet, lurching to help a toddler who slips on the slide, or feeling that tug on your heartstrings for a refugee in a distant land, you will consider this "altruistic urge"—which is by no means perfect, but is natural, adaptive, rational, and sometimes funny or even beautiful.

NOTES

PREFACE

1. See Helen Sullivan, "Florida Man Rescues Puppy from Jaws of Alligator Without Dropping Cigar," *The Guardian*, November 23, 2020, https://www.theguardian.com/us-news/2020/nov/23/man-rescues -puppy-from-alligator-without-dropping-cigar; Vaiva Vareikaite, "60 Times Florida Man Did Something So Crazy We Had to Read the Headings Twice," *Bored Panda* (2018), https://www.boredpanda.com /hilarious-florida-man-headings/?utm_source=google&utm_medium =organic&utm_campaign=organic.
2. "Pregnant Woman Rescues Husband from Shark Attack in Florida," *BBC News*, September 24, 2020, https://www.bbc.com/news/world-us -canada-5428069; Cara Buckley, "Man Is Rescued by Stranger on Subway Tracks," *New York Times*, January 3, 2007.
3. Associated Press, "Dog Dies After Saving Trinidad Man from Fire," *Los Angeles Times*, October 11, 2008.
4. Eleanor Rosch, "Principles of Categorization," in *Cognition and Categorization*, ed. Eleanor Rosch and Barbara B. Lloyd, 27–48 (Hillsdale, NJ: Erlbaum, 1978).

INTRODUCTION

1. William E. Wilsoncroft, "Babies by Bar-Press: Maternal Behavior in the Rat," *Behavior Research Methods, Instruments and Computers* 1 (1969): 229–30, at 229.

2. K. D. Broad, J. P. Curley, and E. B. Keverne, "Mother-Infant Bonding and the Evolution of Mammalian Social Relationships," *Philosophical Transactions of the Royal Society of London. Series B: Biological Sciences* 361, no. 1476 (December 29, 2006): 2199–214. https://doi.org/10.1098/rstb.2006.1940.

3. We use the term "nonhuman animals" in science, even though most people simply say "animals" in contrast to "humans," because humans are animals too. This phrase may seem unnecessarily burdensome, but we need the constant reminder that humans are animals and share much with other species.

1. THE ALTRUISTIC RESPONSE MODEL

1. Scientists use the term "homology" or "homologous" to describe situations where two body structures look similar in two different species because they share a common ancestor, and not just as a coincidence whereby the structure arose independently in two distantly related species. We will talk more about this in the next chapter.

2. Frans B. M. de Waal and Filippo Aureli, "Consolation, Reconciliation, and a Possible Cognitive Difference between Macaque and Chimpanzee," in *Reaching Into Thought: The Minds of the Great Apes*, ed. K. A. Bard, A. E. Russon, and S. T. Parker (Cambridge: Cambridge University Press, 1996), 80–110.

3. John Bowlby, "The Nature of the Child's Tie to His Mother," *International Journal of Psycho-Analysis* 39 (1958): 350–73; Sarah Blaffer Hrdy, *Mothers and Others* (Cambridge, MA: Harvard University Press, 2009); Frans B. M. de Waal, *Good Natured: The Origins of Right and Wrong in Humans and Other Animals* (Cambridge, MA: Harvard University Press, 1996); Stephanie D. Preston, "The Origins of Altruism in Offspring Care," *Psychological Bulletin* 139, no. 6 (2013): 1305–41, https://doi.org/10.1037/a0031755; Stephanie L. Brown, R. Michael Brown, and Louis A. Penner, *Moving Beyond Self-Interest: Perspectives from Evolutionary Biology, Neuroscience, and the Social Sciences* (New York: Oxford University Press, 2011); Abigail A. Marsh, "Neural, Cognitive, and Evolutionary Foundations of Human Altruism," *Wiley Interdisciplinary Reviews: Cognitive Science* 7, no. 1 (2016): 59–71; C. Daniel Batson, "The

Naked Emperor: Seeking a More Plausible Genetic Basis for Psychological Altruism," *Economics and Philosophy* 26, no. 2 (2010): 149–64, https://doi.org/10.1017/S0266267110000179.

4. Michael Numan and Thomas R. Insel, *The Neurobiology of Parental Behavior* (New York: Springer, 2003).

5. Michael Numan, "Neural Circuits Regulating Maternal Behavior: Implications for Understanding the Neural Basis of Social Cooperation and Competition," in Brown, Brown, and Penner, *Moving Beyond Self-Interest*, 89–108.

6. Numan and Insel, *The Neurobiology of Parental Behavior*, 2003; Joseph S. Lonstein and Joan I. Morrell, "Neuroendocrinology and Neurochemistry of Maternal Motivation and Behavior," in *Handbook of Neurochemistry and Molecular Neurobiology*, ed. Abel Lajtha and Jeffrey D. Blaustein, 3rd ed. (Berlin: Springer-Verlag, 2007), 195–245, http://www.springerlink.com/content/nw8357tv143w4w21/; Joseph S. Lonstein and Alison S. Fleming, "Parental Behaviors in Rats and Mice," *Current Protocols in Neuroscience* 15 (2002): Unit 8.15; Jill B. Becker and Jane R. Taylor, "Sex Differences in Motivation," in *Sex Differences in the Brain: From Genes to Behavior*, ed. Jill B. Becker et al. (New York: Oxford University Press, 2008).

7. Michael Numan, "Hypothalamic Neural Circuits Regulating Maternal Responsiveness Toward Infants," *Behavioral & Cognitive Neuroscience Reviews* 5, no. 4 (December 2006): 163–90.

8. Jessika Golle et al., "Sweet Puppies and Cute Babies: Perceptual Adaptation to Babyfacedness Transfers Across Species," *PLoS ONE* 8, no. 3 (March 13, 2013): e58248, https://doi.org/10.1371/journal.pone.0058248.

9. Johan N. Lundström et al., "Maternal Status Regulates Cortical Responses to the Body Odor of Newborns," *Frontiers in Psychology* 4 (September 5, 2013), https://doi.org/10.3389/fpsyg.2013.00597.

10. Note that when we refer to a pup or some chocolate as a "reward," we are not referring to a concrete compensation for "good deeds" but the reinforcing quality attached to any item that motivates us to approach again, which may not be experienced consciously.

11. Stephanie D. Preston, Morten L. Kringelbach, and Brian Knutson, eds., *The Interdisciplinary Science of Consumption* (Cambridge, MA: MIT Press, 2014).

12. Stephanie D. Preston and Andrew D. MacMillan-Ladd, "Object Attachment and Decision-Making," *Current Opinion in Psychology* 39 (June 2021): 31–37, https://doi.org/10.1016/j.copsyc.2020.07.019.

13. Diane S. Berry and Leslie Z. McArthur, "Some Components and Consequences of a Babyface," *Journal of Personality and Social Psychology* 48, no. 2 (1985): 312–23, https://doi.org/10.1037/0022-3514.48.2.312; Leslie A. Zebrowitz, Karen Olson, and Karen Hoffman, "Stability of Babyfaceness and Attractiveness Across the Life Span," *Journal of Personality and Social Psychology* 64, no. 3 (1993): 453–66, https://doi.org/10.1037/0022 -3514.64.3.453.

14. Simon Baron-Cohen and Sally Wheelwright, "The Empathy Quotient: An Investigation of Adults with Asperger Syndrome or High Functioning Autism, and Normal Sex Differences," *Journal of Autism and Developmental Disorders* 34, no. 2 (April 2004): 163–75, https://doi.org /10.1023/B:JADD.0000022607.19833.00; Todd A. Mooradian, Mark Davis, and Kurt Matzler, "Dispositional Empathy and the Hierarchical Structure of Personality," *American Journal of Psychology* 124, no. 1 (2011): 99, https://doi.org/10.5406/amerjpsyc.124.1.0099; Ervin Staub et al., eds., *Development and Maintenance of Prosocial Behavior* (Boston: Springer, 1984), https://doi.org/10.1007/978-1-4613-2645-8.

15. David Hume, *A Treatise of Human Nature* (North Chelmsford, MA: Courier Corporation, 2003); William McDougall, *An Introduction to Social Psychology* (London: Methuen, 1908); John Bowlby, *Attachment and Loss*, vol. 1, *Attachment* (New York: Basic Books, 1969); Charles Darwin, *The Expression of the Emotions in Man and Animals*, 3rd ed. (Oxford: Oxford University Press, 1872); Irenaus Eibl-Eibesfeldt, *Love and Hate*, trans. Geoffrey Strachan, 2nd ed. (New York: Schocken Books, 1971).

16. De Waal, *Good Natured*; Hrdy, *Mothers and Others*; Marsh, "Neural, Cognitive, and Evolutionary Foundations of Human Altruism."

17. Brown, Brown, and Penner, *Moving Beyond Self-Interest*.

18. Numan, "Neural Circuits Regulating Maternal Behavior."

19. Pat Barclay, "Altruism as a Courtship Display: Some Effects of Third-Party Generosity on Audience Perceptions," *British Journal of Psychology* 101, no. 1 (2010): 123–35, https://doi.org/10.1348/000712609X435733.

20. Selwyn W. Becker and Alice H. Eagly, "The Heroism of Women and Men," *American Psychologist* 59, no. 3 (2004): 163–78, https://doi.org/10

.1037/0003-066x.59.3.163; S. P. Oliner, "Extraordinary Acts of Ordinary People," in *Altruism and Altruistic Love: Science, Philosophy, and Religion in Dialogue*, ed. Steven Post et al. (Oxford: Oxford University Press, 2002), 123–39.

2. SIMILARITIES BETWEEN OFFSPRING CARE AND ALTRUISM ACROSS SPECIES

1. Ernst Fehr and Urs Fischbacher, "The Nature of Human Altruism," *Nature* 425, no. 6960 (October 23, 2003): 785–91.
2. Stephanie D. Preston, "The Evolution and Neurobiology of Heroism," in *The Handbook of Heroism and Heroic Leadership*, ed. S. T. Allison, G. R. Goethals, and R. M. Kramer (New York: Taylor & Francis/ Routledge, 2016).
3. William E. Wilsoncroft, "Babies by Bar-Press: Maternal Behavior in the Rat," *Behavior Research Methods, Instruments and Computers* 1 (1969): 229–30.
4. Thomas R. Insel, Stephanie D. Preston, and James T. Winslow, "Mating in the Monogamous Male: Behavioral Consequences," *Physiology & Behavior* 57, no. 4 (1995): 615–27.
5. F. Aureli, S. D. Preston, and F. B. de Waal, "Heart Rate Responses to Social Interactions in Free-Moving Rhesus Macaques (Macaca Mulatta): A Pilot Study," *Journal of Comparative Psychology* 113, no. 1 (March 1999): 59–65.
6. Katherine E. Wynne-Edwards, "Hormonal Changes in Mammalian Fathers," *Hormones and Behavior* 40, no. 2 (September 2001): 139–45, https://doi.org/10.1006/hbeh.2001.1699.
7. Geoffrey Schoenbaum, Andrea A. Chiba, and Michela Gallagher, "Orbitofrontal Cortex and Basolateral Amygdala Encode Expected Outcomes During Learning," *Nature Neuroscience* 1, no. 2 (June 1998): 155–59.
8. R. Nowak et al., "Perinatal Visceral Events and Brain Mechanisms Involved in the Development of Mother-Young Bonding in Sheep," *Hormones and Behavior* 52, no. 1 (2007): 92–98.
9. Trevor W. Robbins, "Homology in Behavioural Pharmacology: An Approach to Animal Models of Human Cognition," *Behavioural Pharmacology* 9, no. 7 (November 1998): 509–19, https://doi.org/10 .1097/00008877-199811000-00005.

10. Dost Öngür and Joseph L. Price, "The Organization of Networks Within the Orbital and Medial Prefrontal Cortex of Rats, Monkeys and Humans," *Cerebral Cortex* 10 (2000): 206–19.

11. James P. Burkett et al., "Oxytocin-Dependent Consolation Behavior in Rodents," *Science* 351, no. 6271 (2016): 375–78.

12. K. Z. Meyza et al., "The Roots of Empathy: Through the Lens of Rodent Models," *Neuroscience & Biobehavioral Reviews* 76 (May 2017): 216–34, https://doi.org/10.1016/j.neubiorev.2016.10.028; Jules B. Panksepp and Garet P. Lahvis, "Rodent Empathy and Affective Neuroscience," *Neuroscience & Biobehavioral Reviews* 35, no. 9 (October 2011): 1864–75, https://doi.org/10.1016/j.neubiorev.2011.05.013.e

13. E. Nowbahari and K. L. Hollis, "Rescue Behavior: Distinguishing between Rescue, Cooperation and Other Forms of Altruistic Behavior," *Communicative & Integrative Biology* 3, no. 2 (2010): 77–79; Katherine Taylor et al., "Precision Rescue Behavior in North American Ants," *Evolutionary Psychology* 11, no. 3 (July 2013): 14747049130100, https://doi.org/10.1177/147470491301100312.

14. Edward O. Wilson, "A Chemical Releaser of Alarm and Digging Behavior in the Ant Pogonomyrmex Badius (Latreille)," *Psyche* 65, no. 2–3 (1958): 41–51.

15. Tony W. Buchanan et al., "The Empathic, Physiological Resonance of Stress," *Social Neuroscience* 7, no. 2 (2012): 1–11, https://doi.org/10.1080/17470919.2011.588723.

16. David C. Knill and Alexandre Pouget, "The Bayesian Brain: The Role of Uncertainty in Neural Coding and Computation," *Trends in Neurosciences* 27, no. 12 (December 2004): 712–19, https://doi.org/10.1016/j.tins.2004.10.007; Joshua I. Gold and Michael N. Shadlen, "Banburismus and the Brain," *Neuron* 36, no. 2 (October 2002): 299–308, https://doi.org/10.1016/S0896-6273(02)00971-6; Ben R. Newell and David R. Shanks, "Unconscious Influences on Decision Making: A Critical Review," *Behavioral and Brain Sciences* 37, no. 1 (February 2014): 1–19, https://doi.org/10.1017/S0140525X12003214.

17. Lori Marino et al., "Relative Volume of the Cerebellum in Dolphins and Comparison with Anthropoid Primates," *Brain, Behavior and Evolution* 56, no. 4 (2000): 204–11, https://doi.org/10.1159/000047205.

18. David F. Sherry, Lucia F. Jacobs, and Steven J. C. Gaulin, "Spatial Memory and Adaptive Specialization of the Hippocampus," *Trends in*

Neurosciences 15, no. 8 (August 1992): 298–303, https://doi.org/10.1016 /0166-2236(92)90080-R.

19. John R. Krebs, "Food-Storing Birds: Adaptive Specialization in Brain and Behaviour?" *Philosophical Transactions of the Royal Society of London. Series B: Biological Sciences* 329, no. 1253 (August 29, 1990): 153–60, https://doi.org/10.1098/rstb.1990.0160.

20. Sara L. Prescott et al., "Enhancer Divergence and Cis-Regulatory Evolution in the Human and Chimp Neural Crest," *Cell* 163, no. 1 (September 2015): 68–83, https://doi.org/10.1016/j.cell.2015.08.036; Douglas H. Erwin and Eric H. Davidson, "The Evolution of Hierarchical Gene Regulatory Networks," *Nature Reviews Genetics* 10, no. 2 (February 2009): 141–48, https://doi.org/10.1038/nrg2499; Sean B. Carroll, "Evo-Devo and an Expanding Evolutionary Synthesis: A Genetic Theory of Morphological Evolution," *Cell* 134, no. 1 (July 2008): 25–36, https://doi.org/10.1016/j.cell.2008.06.030.

21. Stephanie D. Preston, Morten Kringelbach, and Brian Knutson, eds., *The Interdisciplinary Science of Consumption* (Cambridge, MA: MIT Press, 2014).

22. Jorge Moll et al., "Human Fronto-Mesolimbic Networks Guide Decisions About Charitable Donation," *Proceedings of the National Academy of Sciences USA* 103, no. 42 (October 17, 2006): 15623–28; Brian D. Vickers et al., "Motor System Engagement in Charitable Giving: The Offspring Care Effect," forthcoming.

23. Michael Numan et al., "Medial Preoptic Area Interactions with the Nucleus Accumbens-Ventral Pallidum Circuit and Maternal Behavior in Rats," *Behavioural Brain Research* 158, no. 1 (March 7, 2005): 53–68.

24. Michael Numan and Thomas R. Insel, *The Neurobiology of Parental Behavior* (New York: Springer, 2003); J. S. Rosenblatt, "Nonhormonal Basis of Maternal Behavior in the Rat," *Science* 156 (1967): 1512–14.

25. Thomas R. Insel and Carroll R. Harbaugh, "Lesions of the Hypothalamic Paraventricular Nucleus Disrupt the Initiation of Maternal Behavior," *Physiology & Behavior* 45 (1989): 1033–41.

26. Insel, Preston, and Winslow, "Mating in the Monogamous Male."

27. Joseph S. Lonstein and Joan I. Morrell, "Neuroendocrinology and Neurochemistry of Maternal Motivation and Behavior," in *Handbook of Neurochemistry and Molecular Neurobiology*, ed. Abel Lajtha and Jeffrey D.

Blaustein, 3rd ed., 195–245 (Berlin: Springer-Verlag, 2007), http://www
.springerlink.com/content/nw8357tv143w4w21/.

28. Rosenblatt, "Nonhormonal Basis of Maternal Behavior in the Rat";
Harold I. Siegel and Jay S. Rosenblatt, "Estrogen-Induced Maternal
Behavior in Hysterectomized-Ovariectomized Virgin Rats," *Physiol-
ogy & Behavior* 14, no. 4 (1975): 465–71; J. S. Rosenblatt and K. Ceus,
"Estrogen Implants in the Medial Preoptic Area Stimulate Maternal
Behavior in Male Rats," *Hormones and Behavior* 33 (1998): 23–30.

29. Paul D. MacLean, "Brain Evolution Relating to Family, Play, and the
Separation Call," *Archives of General Psychiatry* 42, no. 4 (1985): 405–17.

30. Wynne-Edwards, "Hormonal Changes in Mammalian Fathers";
Toni E. Ziegler, "Hormones Associated with Non-Maternal Infant
Care: A Review of Mammalian and Avian Studies," *Folia Primatolog-
ica* 71, no. 1–2 (2000): 6–21; Wendy Saltzman and Toni E. Ziegler,
"Functional Significance of Hormonal Changes in Mammalian
Fathers," *Journal of Neuroendocrinology* 26, no. 10 (October 2014): 685–96,
https://doi.org/10.1111/jne.12176.

31. Bruce Waldman, "The Ecology of Kin Recognition," *Annual Review of
Ecology and Systematics* 19, no. 1 (November 1988): 543–71, https://doi.org
/10.1146/annurev.es.19.110188.002551; Peter G. Hepper, "Kin Recognition:
Functions and Mechanisms, a Review," *Biological Reviews* 61, no. 1 (February 1986): 63–93, https://doi.org/10.1111/j.1469-185X.1986.tb00427.x.

32. K. M. Kendrick et al., "Neural Control of Maternal Behaviour and
Olfactory Recognition of Offspring," *Brain Research Bulletin* 44, no. 4
(1997): 383–95.

33. Sarah Blaffer Hrdy, *Mothers and Others* (Cambridge, MA: Harvard
University Press, 2009).

34. Toni E. Ziegler and Charles T. Snowdon, "The Endocrinology of Fam-
ily Relationships in Biparental Monkeys," In *The Endocrinology of Social
Relationships*, ed. Peter T. Ellison and Peter B. Gray (Cambridge, MA:
Harvard University Press, 2009), 138–58.

35. Alison S. Fleming et al., "Testosterone and Prolactin Are Associated
with Emotional Responses to Infant Cries in New Fathers," *Hormones
and Behavior* 42, no. 4 (2002): 399–413; Katherine E. Wynne-Edwards
and Mary E. Timonin, "Paternal Care in Rodents: Weakening Support for Hormonal Regulation of the Transition to Behavioral Fatherhood in Rodent Animal Models of Biparental Care," *Hormones and*

Behavior 52, no. 1 (2007): 114–21; Sari M. van Anders, Richard M. Tolman, and Brenda L. Volling, "Baby Cries and Nurturance Affect Testosterone in Men," *Hormones and Behavior* 61, no. 1 (2012): 31–36, https://doi.org/10.1016/j.yhbeh.2011.09.012; James K. Rilling and Jennifer S. Mascaro, "The Neurobiology of Fatherhood," *Current Opinion in Psychology* 15 (June 2017): 26–32, https://doi.org/10.1016/j.copsyc.2017.02.013.

36. A. E. Storey et al., "Hormonal Correlates of Paternal Responsiveness in New and Expectant Fathers," *Evolution and Human Behavior* 21 (2000): 79–95.

37. Fleming et al., "Testosterone and Prolactin Are Associated with Emotional Responses."

38. Van Anders, Tolman, and Volling, "Baby Cries and Nurturance Affect Testosterone in Men."

39. Ronald C. Johnson, "Attributes of Carnegie Medalists Performing Acts of Heroism and of the Recipients of These Acts," *Ethology and Sociobiology* 17, no. 5 (September 1996): 355–62.

40. Margo Wilson, Martin Daly, and Nicholas Pound, "An Evolutionary Psychological Perspective on the Modulation of Competitive Confrontation and Risk-Taking," *Hormones, Brain and Behavior* 5 (2002): 381–408.

41. Rufus A. Johnstone, "Sexual Selection, Honest Advertisement and the Handicap Principle: Reviewing the Evidence," *Biological Reviews* 70 (1995): 1–65.

42. Cara Buckley, "Man Is Rescued by Stranger on Subway Tracks," *New York Times*, January 3, 2007.

43. Robert Seyfarth, Dorothy L. Cheney, and Peter Marler, "Monkey Responses to Three Different Alarm Calls: Evidence of Predator Classification and Semantic Communication," *Science* 210, no. 4471 (November 14, 1980): 801–3, https://doi.org/10.1126/science.7433999.

44. Wynne-Edwards and Timonin, "Paternal Care in Rodents."

45. Dario Maestripieri and Julia L. Zehr, "Maternal Responsiveness Increases During Pregnancy and After Estrogen Treatment in Macaques," *Hormones and Behavior* 34, no. 3 (1998): 223–30, https://doi.org/10.1006/hbeh.1998.1470.

46. Felix Warneken and Michael Tomasello, "Varieties of Altruism in Children and Chimpanzees," *Trends in Cognitive Sciences* 13, no. 9 (2009): 397–402.

47. Hrdy, *Mothers and Others*.

48. Vickers et al., "Motor System Engagement in Charitable Giving"

3. DIFFERENT KINDS OF ALTRUISM

1. Felix Warneken and Michael Tomasello, "Varieties of Altruism in Children and Chimpanzees," *Trends in Cognitive Sciences* 13, no. 9 (2009): 397–402.

2. David J. Hauser, Stephanie D. Preston, and R. Brent Stansfield, "Altruism in the Wild: When Affiliative Motives to Help Positive People Overtake Empathic Motives to Help the Distressed," *Journal of Experimental Psychology: General* 143, no. 3 (December 23, 2014): 1295–1305, https://doi.org/10.1037/a0035464.

3. Frans B. M. de Waal, "Putting the Altruism Back Into Altruism: The Evolution of Empathy," *Annual Review of Psychology* 59 (2008): 279–300.

4. Kristen A. Dunfield, "A Construct Divided: Prosocial Behavior as Helping, Sharing, and Comforting Subtypes," *Frontiers in Psychology* 5 (September 2, 2014): 958, https://doi.org/10.3389/fpsyg.2014.00958.

5. Katherine Taylor et al., "Precision Rescue Behavior in North American Ants," *Evolutionary Psychology* 11, no. 3 (July 2013): 14747049130110. https://doi.org/10.1177/147470491301100312.

6. Anne M. McGuire, "Helping Behaviors in the Natural Environment: Dimensions and Correlates of Helping," *Personality and Social Psychology Bulletin* 20, no. 1 (February 1994): 45–56, https://doi.org/10.1177/0146167294201004.

7. C. Daniel Batson, "Altruism and Prosocial Behavior," in *The Handbook of Social Psychology*, ed. Daniel T. Gilbert, Susan T. Fiske, and Gardner Lindzey, 4th ed. (New York: Oxford University Press, 1998), 2:282–316.

8. Michael Numan and Thomas R. Insel, *The Neurobiology of Parental Behavior* (New York: Springer, 2003).

9. B. R. Vickers et al., "Motor System Engagement in Charitable Giving: The Offspring Care Effect," forthcoming.

10. Frans B. M. de Waal, *Good Natured: The Origins of Right and Wrong in Humans and Other Animals* (Cambridge, MA: Harvard University Press, 1996); Sarah Blaffer Hrdy, *Mothers and Others* (Cambridge, MA: Harvard University Press, 2009); Abigail A. Marsh, "Neural,

Cognitive, and Evolutionary Foundations of Human Altruism," *Wiley Interdisciplinary Reviews: Cognitive Science* 7, no. 1 (2016): 59–71; Debra M. Zeifman, "An Ethological Analysis of Human Infant Crying: Answering Tinbergen's Four Questions," *Developmental Psychobiology* 39, no. 4 (December 2001): 265–85, https://doi.org/10.1002/dev.1005.

11. Frans B. M. de Waal and Filippo Aureli, "Consolation, Reconciliation, and a Possible Cognitive Difference between Macaque and Chimpanzee," in *Reaching Into Thought: The Minds of the Great Apes*, ed. K. A. Bard, A. E. Russon, and S. T. Parker (Cambridge: Cambridge University Press, 1996), 80–110.

12. Sanjida M. O'Connell, "Empathy in Chimpanzees: Evidence for Theory of Mind?" *Primates* 36, no. 3 (1995): 397–410.

13. Orlaith N. Fraser and Thomas Bugnyar, "Do Ravens Show Consolation? Responses to Distressed Others," *PLoS ONE* 5, no. 5 (May 12, 2010): e10605, https://doi.org/10.1371/journal.pone.0010605.

14. Jennifer Crocker, Amy Canevello, and Ashley A. Brown, "Social Motivation: Costs and Benefits of Selfishness and Otherishness," *Annual Review of Psychology* 68, no. 1 (January 3, 2017): 299–325, https://doi .org/10.1146/annurev-psych-010416-044145.

15. Carolyn Zahn-Waxler, Marian Radke-Yarrow, and Robert A. King, "Child Rearing and Children's Prosocial Initiations Toward Victims of Distress," *Child Development* 50, no. 2 (1979): 319–30.

16. Filippo Aureli, Stephanie D. Preston, and Frans B. M. de Waal, "Heart Rate Responses to Social Interactions in Free-Moving Rhesus Macaques (Macaca Mulatta): A Pilot Study," *Journal of Comparative Psychology* 113, no. 1 (March 1999): 59–65; Robert M. Sapolsky, "Stress, Glucocorticoids, and Damage to the Nervous System: The Current State of Confusion," *Stress* 1, no. 1 (2009): 1–19, https://doi .org/10.3109/10253899609001092.

17. De Waal, *Good Natured*.

18. Stephanie L. Brown and R. Michael Brown, "Connecting Prosocial Behavior to Improved Physical Health: Contributions from the Neurobiology of Parenting," *Neuroscience & Biobehavioral Reviews* 55 (August 2015): 1–17, https://doi.org/10.1016/j.neubiorev.2015.04.004.

19. Laura L. Carstensen, John M. Gottman, and Robert W. Levenson, "Emotional Behavior in Long-Term Marriage," *Psychology and Aging* 10, no. 1 (1995): 140–49, https://doi.org/10.1037/0882-7974.10.1.140.

20. Line S. Loken et al., "Coding of Pleasant Touch by Unmyelinated Afferents in Humans," *Nature Neuroscience* 12, no. 5 (2009): 547–48.

21. S. P. Oliner, "Extraordinary Acts of Ordinary People," in *Altruism and Altruistic Love: Science, Philosophy, and Religion in Dialogue*, ed. Steven Post, Lynn G. Underwood, Jeffrey P. Schloss, and William B. Hurlburt (Oxford: Oxford University Press, 2002), 123–39; Zsolt Keczer et al., "Social Representations of Hero and Everyday Hero: A Network Study from Representative Samples," *PLOS ONE* 11, no. 8 (August 15, 2016): e0159354. https://doi.org/10.1371/journal.pone.0159354.

22. John M. Darley and Bibb Latané, "Bystander Intervention in Emergencies: Diffusion of Responsibility," *Journal of Personality and Social Psychology* 8, no. 4 (1968): 377–83.

4. WHAT IS AN INSTINCT?

1. Naheed Rajwani, "Study: Rats Are Nice to One Another," *Chicago Tribune*, January 15, 2014.

2. Beth Azar, "Nature, Nurture: Not Mutually Exclusive," *APA Monitor* 28 (1997): 1–28.

3. Kaiping Peng and Richard E. Nisbett, "Culture, Dialectics, and Reasoning About Contradiction," *American Psychologist* 54, no. 9 (1999): 741–54, https://doi.org/10.1037/0003-066X.54.9.741.

4. Jean Marc Gaspard Itard and François Dagognet, *Victor de l'Aveyron* (Paris: Editions Allia, 1994).

5. Susan Curtiss and Harry A Whitaker, *Genie: A Psycholinguistic Study of a Modern-Day Wild Child* (St. Louis, MO: Elsevier Science, 2014).

6. A. Troisi et al., "Severity of Early Separation and Later Abusive Mothering in Monkeys: What Is the Pathogenic Threshold?" *Journal of Child Psychology and Psychiatry* 30, no. 2 (March 1989): 277–84; Dario Maestripieri, "The Biology of Human Parenting: Insights from Nonhuman Primates," *Neuroscience & Biobehavioral Reviews* 23, no. 3 (1999): 411–22, https://doi.org/10.1016/S0149-7634(98)00042-6.

7. Frances A. Champagne, "Epigenetic Mechanisms and the Transgenerational Effects of Maternal Care," *Frontiers in Neuroendocrinology* 29 (2008): 386–97; Frances A. Champagne et al., "Variations in Maternal

Care in the Rat as a Mediating Influence for the Effects of Environment on Development," *Physiology & Behavior* 79, no. 3 (2003): 359–71.

8. William McDougall, *An Introduction to Social Psychology* (London: Methuen, 1908), 77.

9. J. David Ligon and D. Brent Burt, "Evolutionary Origins," in *Ecology and Evolution of Cooperative Breeding in Birds*, ed. Walter D. Koenig and Janis L. Dickinson (Cambridge: Cambridge University Press, 2004), 5–34.

10. Konrad Lorenz and Nikolaas Tinbergen, "Taxis and Instinkhandlung in der Eirollbewegung der Graugans," *Zeitfrist für Tierpsychologie* 2 (1938): 1–29.

11. James L. Gould, *Ethology: The Mechanisms and Evolution of Behavior* (New York: Norton, 1982), 36.

12. The term "supernormal stimuli" was coined by these early ethologists to refer to objects that possess important properties of the object that naturally elicits a response, but in an extreme way. For example, if the roundness of eggs is important for eliciting a response, a very large and very rounded object would be supernormal and could elicit an event stronger or faster response than a typical egg.

13. Gabriela Lichtenstein, "Selfish Begging by Screaming Cowbirds, a Mimetic Brood Parasite of the Bay-Winged Cowbird," *Animal Behaviour* 61, no. 6 (2001): 1151–58.

14. Burton M. Slotnick, "Disturbances of Maternal Behavior in the Rat Following Lesions of the Cingulate Cortex," *Behaviour* 29, no. 2 (1967): 204–36.

15. Slotnick, "Disturbances of Maternal Behavior."

16. Howard Moltz, "Contemporary Instinct Theory and the Fixed Action Pattern," *Psychological Review* 72, no. 1 (1965): 27–47, https://doi.org/10.1037/h0020275.

17. Slotnick, "Disturbances of Maternal Behavior."

18. Joan E. Strassmann, Yong Zhu, and David C. Queller, "Altruism and Social Cheating in the Social Amoeba Dictyostelium Discoideum," *Nature* 408 (2000): 965–67.

19. Stephanie D. Preston and F. B. M. de Waal, "Empathy: Its Ultimate and Proximate Bases," *Behavioral and Brain Sciences* 25, no. 1 (2002): 1–71, https://doi.org/10.1017/S0140525X02000018.

20. Richard Dawkins, *The Selfish Gene* (Oxford: Oxford University Press, 1976), vii.

21. Stephanie D. Preston, "The Origins of Altruism in Offspring Care," *Psychological Bulletin* 139, no. 6 (2013): 1305–41, https://doi.org/10.1037/a0031755.

22. Joseph S. Lonstein and Joan I. Morrell, "Neuroendocrinology and Neurochemistry of Maternal Motivation and Behavior," in *Handbook of Neurochemistry and Molecular Neurobiology*, ed. Abel Lajtha and Jeffrey D. Blaustein, 3rd ed. (Berlin: Springer-Verlag, 2007), 195–245, http://www.springerlink.com/content/nw8357tv143w4w21/.

23. B. J. Mattson et al., "Comparison of Two Positive Reinforcing Stimuli: Pups and Cocaine Throughout the Postpartum Period," *Behavioral Neuroscience* 115 (2001): 683–94; B. J. Mattson et al., "Preferences for Cocaine or Pup-Associated Chambers Differentiates Otherwise Behaviorally Identical Postpartum Maternal Rats," *Psychopharmacology* 167 (2003): 1–8.

24. Paul Bloom, *Against Empathy: The Case for Rational Compassion* (New York: Ecco, 2017).

25. Sarah Blaffer Hrdy, *Mothers and Others* (Cambridge, MA: Harvard University Press, 2009).

26. John Bowlby, *Attachment and Loss*, vol. 1, *Attachment* (New York: Basic Books, 1969); Christine Acebo and Evelyn B. Thoman, "Role of Infant Crying in the Early Mother-Infant Dialogue," *Physiology & Behavior* 57, no. 3 (1995): 541–47; Preston and de Waal, "Empathy"; Dorothy Einon and Michael Potegal, "Temper Tantrums in Young Children," in *The Dynamics of Aggression: Biological and Social Processes in Dyads and Groups*, ed. Michael Potegal and John F. Knutson (New York: Psychology Press, 1994), 157–94.

27. Shelley E. Taylor et al., "Biobehavioral Responses to Stress in Females: Tend-and-Befriend, Not Fight-or-Flight," *Psychological Review* 107, no. 3 (2000): 411–29.

28. Martin L. Hoffman, "Is Altruism Part of Human Nature?" *Journal of Personality and Social Psychology* 40 (1981): 121–37.

29. Kelly A. Brennan and Phillip R. Shaver, "Dimensions of Adult Attachment, Affect Regulation, and Romantic Relationship Functioning," *Personality and Social Psychology Bulletin* 21, no. 3 (1995): 267–83; Carole M. Pistole, "Adult Attachment Styles: Some Thoughts on

Closeness-Distance Struggles," *Family Process* 33, no. 2 (1994): 147–59, https://doi.org/10.1111/j.1545-5300.1994.00147.x.

30. Stephanie D. Preston, Alicia J. Hofelich, and R. Brent Stansfield, "The Ethology of Empathy: A Taxonomy of Real-World Targets of Need and Their Effect on Observers," *Frontiers in Human Neuroscience* 7, no. 488 (2013): 1–13, https://doi.org/10.3389/fnhum.2013.00488.

31. Padma Kaul et al., "Temporal Trends in Patient and Treatment Delay Among Men and Women Presenting with ST-Elevation Myocardial Infarction," *American Heart Journal* 161, no. 1 (January 2011): 91–97, https://doi.org/10.1016/j.ahj.2010.09.016; Matthew Liakos and Puja B. Parikh, "Gender Disparities in Presentation, Management, and Outcomes of Acute Myocardial Infarction," *Current Cardiology Reports* 20, no. 8 (August 2018): 64, https://doi.org/10.1007/s11886-018-1006-7.

32. Sheila Marikar, "Natasha Richardson Died of Epidural Hematoma After Skiing Accident," ABC News, March 19, 2009.

33. Robyn J. Meyer, Andreas A. Theodorou, and Robert A. Berg, "Childhood Drowning," *Pediatrics in Review* 27, no. 5 (May 2006): 163–69, https://doi.org/10.1542/pir.27-5-163.

34. Daniel Kahneman, *Thinking, Fast and Slow* (New York: Farrar, Straus and Giroux, 2011).

35. Elsa Addessi et al., "Specific Social Influences on the Acceptance of Novel Foods in 2–5-Year-Old Children," *Appetite* 45, no. 3 (December 2005): 264–71, https://doi.org/10.1016/j.appet.2005.07.007; Elisabetta Visalberghi and Elsa Addessi, "Seeing Group Members Eating a Familiar Food Enhances the Acceptance of Novel Foods in Capuchin Monkeys," *Animal Behaviour* 60, no. 1 (July 2000): 69–76, https://doi.org/10.1006/anbe.2000.1425.

36. John M. Darley and Bibb Latané, "Bystander Intervention in Emergencies: Diffusion of Responsibility," *Journal of Personality and Social Psychology* 8, no. 4 (1968): 377–83.

37. Spencer K. Lynn et al., "Decision Making from Economic and Signal Detection Perspectives: Development of an Integrated Framework," *Frontiers in Psychology* 6 (July 8, 2015), https://doi.org/10.3389/fpsyg.2015.00952.

38. John A. Swets, *Signal Detection Theory and ROC Analysis in Psychology and Diagnostics Collected Papers* (New York: Psychology Press, 2014).

39. Robert M. Sapolsky, "The Influence of Social Hierarchy on Primate Health," *Science* 308, no. 5722 (April 29, 2005): 648–52, https://doi.org/10.1126/science.1106477.

40. Lori L. Heise, "Violence Against Women: An Integrated, Ecological Framework," *Violence Against Women* 4, no. 3 (June 1998): 262–90, https://doi.org/10.1177/1077801298004003002.

41. Wolfram Schultz, "Neural Coding of Basic Reward Terms of Animal Learning Theory, Game Theory, Microeconomics and Behavioural Ecology," *Current Opinion in Neurobiology* 14, no. 2 (April 2004): 139–47, https://doi.org/10.1016/j.conb.2004.03.017.

42. Sapolsky, "The Influence of Social Hierarchy on Primate Health."

43. Björn Brembs and Jan Wiener, "Context and Occasion Setting in Drosophila Visual Learning," *Learning & Memory* 13, no. 5 (September 1, 2006): 618–28, https://doi.org/10.1101/lm.318606; Kurt Gray, Adrian F. Ward, and Michael I. Norton, "Paying It Forward: Generalized Reciprocity and the Limits of Generosity," *Journal of Experimental Psychology: General* 143, no. 1 (2014): 247–54, https://doi.org/10.1037/a0031047; David DeSteno et al., "Gratitude as Moral Sentiment: Emotion-Guided Cooperation in Economic Exchange," *Emotion* 10, no. 2 (2010): 289–93, https://doi.org/10.1037/a0017883; Lalin Anik et al., "Feeling Good About Giving: The Benefits (and Costs) of Self-Interested Charitable Behavior," *SSRN Electronic Journal* 2009, https://doi.org/10.2139/ssrn.1444831.

5. THE NEURAL BASES OF ALTRUISM

1. Stephanie D. Preston, "The Origins of Altruism in Offspring Care," *Psychological Bulletin* 139, no. 6 (2013): 1305–41, https://doi.org/10.1037/a0031755.

2. Theodore C. Schneirla, "An Evolutionary and Developmental Theory of Biphasic Processes Underlying Approach and Withdrawal," *Nebraska Symposium on Motivation* 7 (1959): 1–42.

3. Alison S. Fleming, Michael Numan, and Robert S. Bridges, "Father of Mothering: Jay S. Rosenblatt," *Hormones and Behavior* 55, no. 4 (April 2009): 484–87, https://doi.org/10.1016/j.yhbeh.2009.01.001.

4. Thomas R. Insel and Larry J. Young, "The Neurobiology of Attachment," *Nature Reviews Neuroscience* 2, no. 2 (February 2001): 129–36.

5. Stephanie D. Preston, Morten Kringelbach, and Brian Knutson, eds., *The Interdisciplinary Science of Consumption* (Cambridge, MA: MIT Press, 2014).

6. William E. Wilsoncroft, "Babies by Bar-Press: Maternal Behavior in the Rat," *Behavior Research Methods, Instruments and Computers* 1 (1969): 229–30.

7. Allan R. Wagner, "Effects of Amount and Percentage of Reinforcement and Number of Acquisition Trials on Conditioning and Extinction," *Journal of Experimental Psychology* 62, no. 3 (1961): 234–42, https://doi.org/10.1037/h0042251; Norman E. Spear, Winfred F. Hill, and Denis J. O'Sullivan, "Acquisition and Extinction after Initial Trials Without Reward," *Journal of Experimental Psychology* 69, no. 1 (1965): 25–29, https://doi.org/10.1037/h0021628.

8. Frédéric Levy, Matthieu Keller, and Pascal Poindron, "Olfactory Regulation of Maternal Behavior in Mammals," *Hormones and Behavior* 46, no. 3 (September 2004): 284–302.

9. Kent C. Berridge and Terry E. Robinson, "What Is the Role of Dopamine in Reward: Hedonic Impact, Reward Learning, or Incentive Salience?" *Brain Research Reviews* 28, no. 3 (December 1998): 309–69.

10. Stefan Hansen, "Maternal Behavior of Female Rats with 6-OHDA Lesions in the Ventral Striatum: Characterization of the Pup Retrieval Deficit," *Physiology & Behavior* 55, no. 4 (1994): 615–20, https://doi.org/10.1016/0031-9384(94)90034-5.

11. Preston, "The Origins of Altruism in Offspring Care."

12. Susana Peciña and Kent C. Berridge, "Hedonic Hot Spot in Nucleus Accumbens Shell: Where Do μ-Opioids Cause Increased Hedonic Impact of Sweetness?" *Journal of Neuroscience* 25, no. 50 (2005): 11777–86.

13. Kevin D. Broad, James P. Curley, and Eric B. Keverne, "Mother-Infant Bonding and the Evolution of Mammalian Social Relationships," *Philosophical Transactions of the Royal Society of London. Series B: Biological Sciences* 361, no. 1476 (December 29, 2006): 2199–214, https://doi.org/10.1098/rstb.2006.1940.

14. Judith M. Stern and Joseph S. Lonstein, "Neural Mediation of Nursing and Related Maternal Behaviors," *Progress in Brain Research* 133 (2001): 263–78.

15. Jennifer R. Brown et al., "A Defect in Nurturing in Mice Lacking the Immediate Early Gene FosB," *Cell* 86, no. 2 (1996): 297–309.

16. C. A. Pedersen et al., "Oxytocin Activates the Postpartum Onset of Rat Maternal Behavior in the Ventral Tegmental and Medial Preoptic Areas," *Behavioral Neuroscience* 108 (1994): 1163–71.

17. Thomas R. Insel and Carroll R. Harbaugh, "Lesions of the Hypothalamic Paraventricular Nucleus Disrupt the Initiation of Maternal Behavior," *Physiology & Behavior* 45 (1989): 1033–41.

18. Pedersen et al., "Oxytocin Activates the Postpartum Onset."

19. Michael Numan and Thomas R. Insel, *The Neurobiology of Parental Behavior* (New York: Springer, 2003); Insel and Young, "The Neurobiology of Attachment."

20. Horst Schulz, Gábor L. Kovács, and Gyula Telegdy, "Action of Posterior Pituitary Neuropeptides on the Nigrostriatal Dopaminergic System," *European Journal of Pharmacology* 57, no. 2–3 (August 1979): 185–90, https://doi.org/10.1016/0014-2999(79)90364-9.

21. M. M. McCarthy, L-M. Kow, and D. W. Pfaff, "Speculations Concerning the Physiological Significance of Central Oxytocin in Maternal Behavior," *Annals of the New York Academy of Sciences* 652 (1992): 70–82, https://doi.org/10.1111/j.1749-6632.1992.tb34347.x.

22. Sarah Blaffer Hrdy, *Mothers and Others* (Cambridge, MA: Harvard University Press, 2009).

23. Charles M. Grinstead and J. Laurie Snell, "Chapter 9: Central Limit Theorem," in *Introduction to Probability*, 2nd ed. (Providence, RI: American Mathematical Society, 1997).

24. Galton also discovered "regression to the mean" here, because children's height was not represented by the same distribution of their parents' heights; rather, the mean reflected that of the population at large and shifted to match each individual only by a proportion of the extremity of their own parents' height.

25. J. Stallings et al., "The Effects of Infant Cries and Odors on Sympathy, Cortisol, and Autonomic Responses in New Mothers and Nonpostpartum Women," *Parenting-Science and Practice* 1, no. 1–2 (2001): 71–100; Alison S. Fleming and Jay S. Rosenblatt, "Olfactory Regulation of Maternal Behavior in Rats: II. Effects of Peripherally Induced Anosmia and Lesions of the Lateral Olfactory Tract in Pup-Induced Virgins," *Journal of Comparative and Physiological Psychology* 86 (1974): 233–46.

26. William O. Beeman, "Making Grown Men Weep," in *Aesthetics in Performance: Formations of Symbolic Instruction and Experience*, ed. Angela

Hobart and Bruce Kapferer (New York: Berghahn Books, 2005), 23–42.

27. Antoine Bechara, Hanna Damasio, and Antonio R. Damasio, "Emotion, Decision Making and the Orbitofrontal Cortex," *Cerebral Cortex* 10, no. 3 (2000): 295–307.

28. John O'Doherty, "Can't Learn Without You: Predictive Value Coding in Orbitofrontal Cortex Requires the Basolateral Amygdala," *Neuron* 39, no. 5 (August 28, 2003): 731–33.

29. A. Bechara et al., "Dissociation of Working Memory from Decision Making Within the Human Prefrontal Cortex," *Journal of Neuroscience* 18 (1998): 428–37; Tina L. Jameson, John M. Hinson, and Paul Whitney, "Components of Working Memory and Somatic Markers in Decision Making," *Psychonomic Bulletin & Review* 11, no. 3 (2004): 515–20; Amy L. Krain et al., "Distinct Neural Mechanisms of Risk and Ambiguity: A Meta-Analysis of Decision-Making," *NeuroImage* 32, no. 1 (2006): 477–84.

30. Bechara et al., "Dissociation of Working Memory from Decision Making within the Human Prefrontal Cortex."

31. Daniel Tranel and Antonio R. Damasio, "The Covert Learning of Affective Valence Does Not Require Structures in Hippocampal System or Amygdala," *Journal of Cognitive Neuroscience* 5, no. 1 (January 1993): 79–88, https://doi.org/10.1162/jocn.1993.5.1.79. This elevated performance for obtaining a food reward may suggest that his intact dopaminergic NAcc processes permitted his intact ability to direct energy toward the "good" provider.

32. Frans B. M. de Waal and Stephanie D. Preston, "Mammalian Empathy: Behavioural Manifestations and Neural Basis," *Nature Reviews Neuroscience* 18, no. 8 (2017): 498–510.

33. Krain et al., "Distinct Neural Mechanisms of Risk and Ambiguity: A Meta-Analysis of Decision-Making."

34. This example also demonstrates the particularly direct link between natural rewards such as food and drink that stimulate the dopaminergic NAcc motivation to act across species and our motivation to act, owing to the way the brain evolved this powerful mechanism to direct behavior.

35. James Andreoni, William T. Harbaugh, and Lise Vesterlund, "Altruism in Experiments," in *The New Palgrave Dictionary of Economics*, ed.

Steven N. Durlauf and Lawrence E. Bloom (London: Palgrave Macmillan, 2008), 134–38.

36. Ernst Fehr and Simon Gächter, "Altruistic Punishment in Humans," *Nature* 415, no. 6868 (January 10, 2002): 137–40; Ernst Fehr and Urs Fischbacher, "The Nature of Human Altruism," *Nature* 425, no. 6960 (October 23, 2003): 785–91; Ernst Fehr and Colin F. Camerer, "Social Neuroeconomics: The Neural Circuitry of Social Preferences," *Trends in Cognitive Sciences* 11, no. 10 (October 2007): 419–27.

37. A. G. Sanfey et al., "The Neural Basis of Economic Decision-Making in the Ultimatum Game," *Science* 300, no. 5626 (June 13, 2003): 1755–58.

38. D. Knoch et al., "Studying the Neurobiology of Social Interaction with Transcranial Direct Current Stimulation—The Example of Punishing Unfairness," *Cerebral Cortex* 18, no. 9 (September 2008): 1987–90; D. Knoch et al., "Diminishing Reciprocal Fairness by Disrupting the Right Prefrontal Cortex," *Science* 314, no. 5800 (November 3, 2006): 829–32.

39. M. Koenigs and D. Tranel, "Irrational Economic Decision-Making After Ventromedial Prefrontal Damage: Evidence from the Ultimatum Game," *Journal of Neuroscience* 27, no. 4 (January 24, 2007): 951–56.

40. K. McCabe et al., "A Functional Imaging Study of Cooperation in Two-Person Reciprocal Exchange," *Proceedings of the National Academy of Sciences USA* 98, no. 20 (September 25, 2001): 11832–35.

41. F. Krueger et al., "Neural Correlates of Trust," *Proceedings of the National Academy of Sciences USA* 104, no. 50 (December 11, 2007): 20084–89.

42. D. J. de Quervain et al., "The Neural Basis of Altruistic Punishment," *Science* 305, no. 5688 (August 27, 2004): 1254–58.

43. J. Rilling et al., "A Neural Basis for Social Cooperation," *Neuron* 35, no. 2 (2002): 395–405.

44. James K. Rilling et al., "Opposing BOLD Responses to Reciprocated and Unreciprocated Altruism in Putative Reward Pathways," *Neuroreport* 15, no. 16 (2004): 2539–43.

45. T. Singer et al., "Empathic Neural Responses Are Modulated by the Perceived Fairness of Others," *Nature* 439, no. 7075 (January 26, 2006): 466–69.

46. Paul J. Zak, "The Neurobiology of Trust," *Scientific American* 298, no. 6 (June 2008): 88–92, 95; Paul J. Zak, Robert Kurzban, and William T.

Matzner, "Oxytocin Is Associated with Human Trustworthiness," *Hormones and Behavior* 48, no. 5 (December 2005): 522–27.

47. M. Kosfeld et al., "Oxytocin Increases Trust in Humans," *Nature* 435, no. 7042 (June 2, 2005): 673–76; Paul J. Zak, Angela A. Stanton, and Sheila Ahmadi, "Oxytocin Increases Generosity in Humans," *PLoS ONE* 2, no. 11 (2007): e1128; Vera B. Morhenn et al., "Monetary Sacrifice Among Strangers Is Mediated by Endogenous Oxytocin Release After Physical Contact," *Evolution and Human Behavior* 29, no. 6 (2008): 375–83.

48. Salomon Israel et al., "Molecular Genetic Studies of the Arginine Vasopressin 1a Receptor (AVPR1a) and the Oxytocin Receptor (OXTR) in Human Behaviour: From Autism to Altruism with Some Notes in Between," *Progress in Brain Research* 170 (2008): 435–49.

49. T. Baumgartner et al., "Oxytocin Shapes the Neural Circuitry of Trust and Trust Adaptation in Humans," *Neuron* 58, no. 4 (May 22, 2008): 639–50.

50. T. Singer et al., "Effects of Oxytocin and Prosocial Behavior on Brain Responses to Direct and Vicariously Experienced Pain," *Emotion* 8, no. 6 (December 2008): 781–91.

51. Marian J. Bakermans-Kranenburg and Marinus H. van IJzendoorn, "A Sociability Gene? Meta-Analysis of Oxytocin Receptor Genotype Effects in Humans," *Psychiatric Genetics* 24, no. 2 (April 2014): 45–51, https://doi.org/10.1097/YPG.0b013e3283643684; Gideon Nave, Colin Camerer, and Michael McCullough, "Does Oxytocin Increase Trust in Humans? A Critical Review of Research," *Perspectives on Psychological Science* 10, no. 6 (November 2015): 772–89, https://doi.org/10.1177/1745691615600138; Marinus H. Van IJzendoorn and Marian J. Bakermans-Kranenburg, "A Sniff of Trust: Meta-Analysis of the Effects of Intranasal Oxytocin Administration on Face Recognition, Trust to in-Group, and Trust to out-Group," *Psychoneuroendocrinology* 37, no. 3 (March 2012): 438–43, https://doi.org/10.1016/j.psyneuen.2011.07.008.

52. Stephanie D. Preston, "The Rewarding Nature of Social Contact," *Science (New York, N.Y.)* 357, no. 6358 (29 2017): 1353–54, https://doi.org/10.1126/science.aao7192.

53. William T. Harbaugh, Ulrich Mayr, and Daniel R. Burghart, "Neural Responses to Taxation and Voluntary Giving Rebel Motives for Charitable Donation," *Science* 316 (2007): 1622–25.

54. J. Moll et al., "Human Fronto-Mesolimbic Networks Guide Decisions About Charitable Donation," *Proceedings of the National Academy of Sciences USA* 103, no. 42 (October 17, 2006): 15623–28.

55. Jeffrey P. Lorberbaum et al., "Feasibility of Using FMRI to Study Mothers Responding to Infant Cries," *Depression and Anxiety* 10, no. 3 (1999): 99–104; Jeffrey P. Lorberbaum et al., "A Potential Role for Thalamocingulate Circuitry in Human Maternal Behavior," *Biological Psychiatry* 51, no. 6 (2002): 431–45; Preston, "The Origins of Altruism in Offspring Care."

56. B. R. Vickers et al., "Motor System Engagement in Charitable Giving: The Offspring Care Effect," forthcoming.

57. Felix Warneken and Michael Tomasello, "Varieties of Altruism in Children and Chimpanzees," *Trends in Cognitive Sciences* 13, no. 9 (2009): 397–402.

6. CHARACTERISTICS OF THE VICTIM
THAT FACILITATE A RESPONSE

1. Sarah Blaffer Hrdy, *Mothers and Others* (Cambridge, MA: Harvard University Press, 2009).

2. Cara Buckley, "Man Is Rescued by Stranger on Subway Tracks," *New York Times*, January 3, 2007.

3. Lisa Farwell and Bernard Weiner, "Bleeding Hearts and the Heartless: Popular Perceptions of Liberal and Conservative Ideologies," *Personality and Social Psychology Bulletin* 26, no. 7 (September 2000): 845–52, https://doi.org/10.1177/0146167200269009.

4. Jason T. Newsom and Richard Schulz, "Caregiving from the Recipient's Perspective: Negative Reactions to Being Helped," *Health Psychology* 17, no. 2 (1998): 172–81, https://doi.org/10.1037/0278-6133.17.2.172.

5. Carmel Bitondo Dyer et al., "The High Prevalence of Depression and Dementia in Elder Abuse or Neglect," *Journal of the American Geriatrics Society* 48, no. 2 (February 2000): 205–8, https://doi.org/10.1111/j.1532-5415.2000.tb03913.x; Karl Pillemer and David W. Moore, "Abuse of Patients in Nursing Homes: Findings from a Survey of Staff," *The Gerontologist* 29, no. 3 (June 1, 1989): 314–20, https://doi.org/10.1093/geront/29.3.314.

6. Shane Frederick, George Loewenstein, and Ted O'Donoghue, "Time Discounting and Time Preference: A Critical Review," *Journal of Economic Literature* 40, no. 2 (June 2002): 351–401, https://doi.org/10.1257/jel.40.2.351.

7. Hal E. Hershfield, Taya R. Cohen, and Leigh Thompson, "Short Horizons and Tempting Situations: Lack of Continuity to Our Future Selves Leads to Unethical Decision Making and Behavior," *Organizational Behavior and Human Decision Processes* 117, no. 2 (March 2012): 298–310, https://doi.org/10.1016/j.obhdp.2011.11.002.

8. M. Shiota et al., "Positive Affect and Behavior Change," *Current Opinion in Behavioral Sciences* 39 (2021): 222–28.

9. Elizabeth W. Dunn, Laura B. Aknin, and Michael I. Norton, "Spending Money on Others Promotes Happiness," *Science* 319, no. 5870 (March 21, 2008): 1687–88, https://doi.org/10.1126/science.1150952.

10. J. Andreoni, "Impure Altruism and Donations to Public Goods: A Theory of Warm-Glow Giving," *The Economic Journal* 100, no. 401 (1990): 464–77.

11. Paul Bloom, *Against Empathy: The Case for Rational Compassion* (New York: Ecco, 2016).

12. Konrad Lorenz, "Die Angeborenen Formen Möglicher Erfahrung [The Innate Forms of Potential Experience]," *Zeitschrift für Tierpsychologie* 5 (1943): 233–519.

13. Wulf Schiefenhövel, *Geburtsverhalten und Reproduktive Strategien der Eipo: Ergebnisse Humanethologischer und Ethnomedizinischer Untersuchungen im Zentralen Bergland von Irian Jaya (West-Neuguinea), Indonesien* [Birth Behavior and Reproductive Strategies of the Eipo: Results of Human Ethology and Ethnomedical Researches in the Central Highlands of Irian Jaya (West New Guinea), Indonesia] (Berlin: D. Reimer, 1988).

14. Hiroshi Nittono et al., "The Power of Kawaii: Viewing Cute Images Promotes a Careful Behavior and Narrows Attentional Focus," *PLoS ONE* 7, no. 9 (September 26, 2012): e46362, https://doi.org/10.1371/journal.pone.0046362.

15. Kana Kuraguchi, Kosuke Taniguchi, and Hiroshi Ashida, "The Impact of Baby Schema on Perceived Attractiveness, Beauty, and Cuteness in Female Adults," *SpringerPlus* 4, no. 1 (December 2015): 164, https://doi.org/10.1186/s40064-015-0940-8.

16. Diane S. Berry and Leslie Z. McArthur, "Some Components and Consequences of a Babyface," *Journal of Personality and Social Psychology* 48, no. 2 (1985): 312–23, https://doi.org/10.1037/0022-3514.48.2.312.

17. Caroline F. Keating et al., "Do Babyfaced Adults Receive More Help? The (Cross-Cultural) Case of the Lost Resume," *Journal of Nonverbal Behavior* 27, no. 2 (2003): 89–109.

18. Linda Qui, "5 Irresistible National Geographic Cover Photos," n.d., https://www.nationalgeographic.com/news/2014/12/141206-magazine-covers-photography-national-geographic-afghan-girl/.

19. Ruth Holliday and Joanna Elfving-Hwang, "Gender, Globalization and Aesthetic Surgery in South Korea," *Body & Society* 18, no. 2 (June 2012): 58–81, https://doi.org/10.1177/1357034X12440828.

20. Abigail A. Marsh and Robert E. Kleck, "The Effects of Fear and Anger Facial Expressions on Approach- and Avoidance-Related Behaviors," *Emotion* 5, no. 1 (2005): 119–24.

21. Duane Quiatt, "Aunts and Mothers: Adaptive Implications of Allomaternal Behavior of Nonhuman Primates," *American Anthropologist* 81, no. 2 (June 1979): 310–19, https://doi.org/10.1525/aa.1979.81.2.02a00040.

22. Deborah A. Small and George Loewenstein, "Helping a Victim or Helping the Victim: Altruism and Identifiability," *Journal of Risk and Uncertainty* 26, no. 1 (2003): 5–16, https://doi.org/10.1023/A:102229942219; Tehila Kogut and Ilana Ritov, "The 'Identified Victim' Effect: An Identified Group, or Just a Single Individual?" *Journal of Behavioral Decision Making* 18, no. 3 (July 2005): 157–67, https://doi.org/10.1002/bdm.492; Karen Jenni and George Loewenstein, "Explaining the Identifiable Victim Effect," *Journal of Risk and Uncertainty* 14, no. 3 (1997): 235–57, https://doi.org/10.1023/A:1007740225484.

23. Stephanie D. Preston et al., "A Case Study of a Conservation Flagship Species: The Monarch Butterfly," *Biodiversity and Conservation* 30 (2021): 2057–77.

24. Hrdy, *Mothers and Others.*

25. Paul D. MacLean, *The Triune Brain in Evolution: Role in Paleocerebral Functions* (New York: Plenum Press, 1990).

26. Rebecca M. Kilner, David G. Noble, and Nicholas B. Davies, "Signals of Need in Parent-Offspring Communication and Their Exploitation by the Common Cuckoo," *Nature* 397, no. 6721 (1999): 667–72.

27. Gabriela Lichtenstein, "Selfish Begging by Screaming Cowbirds, a Mimetic Brood Parasite of the Bay-Winged Cowbird," *Animal Behaviour* 61, no. 6 (2001): 1151–58.

28. Susan D. Healy, Selvino R. Dekort, and Nicola S. Clayton, "The Hippocampus, Spatial Memory and Food Hoarding: A Puzzle Revisited," *Trends in Ecology & Evolution* 20, no. 1 (January 2005): 17–22, https://doi.org/10.1016/j.tree.2004.10.006.

29. D. F. Sherry et al., "Females Have a Larger Hippocampus Than Males in the Brood-Parasitic Brown-Headed Cowbird," *Proceedings of the National Academy of Sciences* 90, no. 16 (August 15, 1993): 7839–43, https://doi.org/10.1073/pnas.90.16.7839.

30. Nicola S. Clayton, Juan C. Reboreda, and Alex Kacelnik, "Seasonal Changes of Hippocampus Volume in Parasitic Cowbirds," *Behavioural Processes* 41, no. 3 (December 1997): 237–43, https://doi.org/10.1016/S0376 -6357(97)00050-8.

31. Abigail A. Marsh, Megan N. Kozak, and Nalini Ambady, "Accurate Identification of Fear Facial Expressions Predicts Prosocial Behavior," *Emotion* 7, no. 2 (2007): 239–51.

32. Myron A. Hofer, "Multiple Regulators of Ultrasonic Vocalization in the Infant Rat," *Psychoneuroendocrinology* 21, no. 2 (February 1996): 203–17, https://doi.org/10.1016/0306-4530(95)00042-9.

33. Gwen E. Gustafson and James A. Green, "On the Importance of Fundamental Frequency and Other Acoustic Features in Cry Perception and Infant Development," *Child Development* 60, no. 4 (1989): 772–80.

34. Harvey Fletcher and W. A. Munson, "Loudness, Its Definition, Measurement and Calculation," *Journal of the Acoustical Society of America* 5 (1933): 82–108.

35. K. Michelsson et al., "Crying in Separated and Non-Separated Newborns: Sound Spectrographic Analysis," *Acta Paediatrica* 85, no. 4 (April 1996): 471–75, https://doi.org/10.1111/j.1651-2227.1996.tb14064.x.

36. James J. Gross and Robert W. Levenson, "Emotion Elicitation Using Films," *Cognition & Emotion* 9, no. 1 (January 1995): 87–108, https://doi.org/10.1080/02699939508408966.

37. Michael Macht and Jochen Mueller, "Immediate Effects of Chocolate on Experimentally Induced Mood States," *Appetite* 49, no. 3 (November 2007): 667–74, https://doi.org/10.1016/j.appet.2007.05.004.

38. Alan R. Wiesenfeld and Rafael Klorman, "The Mother's Psychophysiological Reactions to Contrasting Affective Expressions by Her Own
and an Unfamiliar Infant," *Developmental Psychology* 14, no. 3 (1978):
294–304, https://doi.org/10.1037/0012-1649.14.3.294.

39. Nancy Eisenberg et al., "The Relations of Emotionality and Regulation to Dispositional and Situational Empathy-Related Responding,"
Journal of Personality & Social Psychology 66, no. 4 (1994): 776–97.

40. Gustafson and Green, "On the Importance of Fundamental Frequency
and Other Acoustic Features."

41. Birgit Mampe et al., "Newborns' Cry Melody Is Shaped by Their
Native Language," *Current Biology* 19, no. 23 (December 2009):
1994–97, https://doi.org/10.1016/j.cub.2009.09.064.

42. Ervin Staub, "A Child in Distress: The Influence of Nurturance and
Modeling on Children's Attempts to Help," *Developmental Psychology*
5, no. 1 (1971): 124–32, https://doi.org/10.1037/h0031084.

43. Heidi Keller and Hiltrud Otto, "The Cultural Socialization of Emotion Regulation During Infancy," *Journal of Cross-Cultural Psychology*
40, no. 6 (November 2009): 996–1011, https://doi.org/10.1177/00220
22109348576.

44. Stephanie D. Preston, Alicia J. Hofelich, and R. Brent Stansfield, "The
Ethology of Empathy: A Taxonomy of Real-World Targets of Need
and Their Effect on Observers," *Frontiers in Human Neuroscience* 7,
no. 488 (2013): 1–13, https://doi.org/10.3389/fnhum.2013.00488.

45. Jamil Zaki, Niall Bolger, and Kevin N. Ochsner, "It Takes Two: The
Interpersonal Nature of Empathic Accuracy," *Psychological Science* 19,
no. 4 (April 2008): 399–404, https://doi.org/10.1111/j.1467-9280.2008
.02099.x.

46. Frans B. M. de Waal and Stephanie D. Preston, "Mammalian Empathy: Behavioural Manifestations and Neural Basis," *Nature Reviews
Neuroscience* 18, no. 8 (2017): 498–510.

47. Hendrik Hertzberg, "Second Those Emotions: Hillary's Tears," *The
New Yorker*, January 21, 2008.

48. C. Daniel Batson and Jay S. Coke, "Empathy: A Source of Altruistic
Motivation for Helping," in *Altruism and Helping Behavior*, ed. J. Philippe
Rushton and Richard M. Sorrentino (Hillsdale, NJ: Erlbaum, 1981).

49. David J. Hauser, Stephanie D. Preston, and R. Brent Stansfield, "Altruism in the Wild: When Affiliative Motives to Help Positive People

Overtake Empathic Motives to Help the Distressed," *Journal of Experimental Psychology: General* 143, no. 3 (December 23, 2014): 1295–1305, https://doi.org/10.1037/a0035464.

50. Michael Potegal and John F. Knutson, *The Dynamics of Aggression: Biological and Social Processes in Dyads and Groups* (Hillsdale, NJ: Erlbaum, 1994).

51. Tony W. Buchanan and Stephanie D. Preston, "Stress Leads to Prosocial Action in Immediate Need Situations," *Frontiers in Behavioral Neuroscience* 8, no. 5 (2014), https://doi.org/10.3389/fnbeh.2014.00005.

52. Robert M. Sapolsky, "Stress, Glucocorticoids, and Damage to the Nervous System: The Current State of Confusion," *Stress* 1, no. 1 (2009): 1–19, https://doi.org/10.3109/10253899609001092.

53. Gerald S. Wilkinson, "Food Sharing in Vampire Bats," *Scientific American* 262 (1990): 76–82.

7. CHARACTERISTICS OF THE OBSERVER THAT FACILITATE A RESPONSE

1. Michael Numan, "Motivational Systems and the Neural Circuitry of Maternal Behavior in the Rat," *Developmental Psychobiology* 49, no. 1 (January 2007): 12–21.

2. John F. Dovidio et al., *The Social Psychology of Prosocial Behavior* (Philadelphia: Erlbaum, 2006).

3. Caroline E. Zsambok and Gary A. Klein, *Naturalistic Decision Making* (Philadelphia: Erlbaum, 1997).

4. Cara Buckley, "Man Is Rescued by Stranger on Subway Tracks," *New York Times*, January 3, 2007.

5. Selwyn W. Becker and Alice H. Eagly, "The Heroism of Women and Men," *American Psychologist* 59, no. 3 (2004): 163–78, https://doi.org/10.1037/0003-066x.59.3.163.

6. William H. Warren, "Perceiving Affordances: Visual Guidance of Stair Climbing," *Journal of Experimental Psychology: Human Perception and Performance* 10, no. 5 (1984): 683–703, https://doi.org/10.1037/0096-1523.10.5.683.

7. Giacomo Rizzolatti et al., "Premotor Cortex and the Recognition of Motor Actions," *Cognitive Brain Research* 3, no. 2 (March 1996): 131–41, https://doi.org/10.1016/0926-6410(95)00038-0.

8. Albert Bandura, "Self-Efficacy," in *The Corsini Encyclopedia of Psychology*, ed. Irving B. Weiner and W. Edward Craighead (Hoboken, NJ: Wiley, 2010), 1–3; corpsy0836, https://doi.org/10.1002/9780470479216 .corpsy0836.

9. Sharon Connell et al., " 'If It Doesn't Directly Affect You, You Don't Think About It': A Qualitative Study of Young People's Environmental Attitudes in Two Australian Cities," *Environmental Education Research* 5, no. 1 (February 1999): 95–113, https://doi.org/10.1080/1350 462990050106.

10. Icek Ajzen, "The Theory of Planned Behavior," *Organizational Behavior and Human Decision Processes* 50, no. 2 (December 1991): 179–211, https://doi.org/10.1016/0749-5978(91)90020-T.

11. Alice Jones, "The Psychology of Sustainability: What Planners Can Learn from Attitude Research," *Journal of Planning Education and Research* 16, no. 1 (September 1996): 56–65, https://doi.org/10.1177 /0739456X9601600107.

12. Paul Slovic, "If I Look at the Mass I Will Never Act: Psychic Numbing and Genocide," in *Emotions and Risky Technologies*, ed. Sabine Roeser (Dordrecht: Springer Netherlands, 2010), 5:37–59, https://doi.org/10 .1007/978-90-481-8647-1_3.

13. Stephan Dickert et al., "Scope Insensitivity: The Limits of Intuitive Valuation of Human Lives in Public Policy," *Journal of Applied Research in Memory and Cognition* 4, no. 3 (2015): 248–55.

14. Deborah A. Small, George Loewenstein, and Paul Slovic, "Sympathy and Callousness: The Impact of Deliberative Thought on Donations to Identifiable and Statistical Victims," *Organizational Behavior and Human Decision Processes* 102, no. 2 (March 2007): 143–53, https://doi .org/10.1016/j.obhdp.2006.01.005.

15. Stephanie D. Preston et al., "A Case Study of a Conservation Flagship Species: The Monarch Butterfly," *Biodiversity and Conservation* 30 (2021): 2057–77.

16. Stephanie D. Preston et al., "Leveraging Differences in How Liberals versus Conservatives Think about the Earth Improves Pro-Environmental Responses," forthcoming.

17. Patrick M. Rooney, "The Growth in Total Household Giving Is Camouflaging a Decline in Giving by Small and Medium Donors: What Can We Do About It?" *Nonprofit Quarterly*, August 27, 2019,

https://nonprofitquarterly.org/total-household-growth-decline-small
-medium-donors/.

18. Paul Bloom, *Against Empathy: The Case for Rational Compassion* (New York: Ecco, 2016).

19. John M. Darley and Bibb Latané, "Bystander Intervention in Emergencies: Diffusion of Responsibility," *Journal of Personality and Social Psychology* 8, no. 4 (1968): 377–83; Bibb Latané and John M. Darley, "Bystander 'Apathy,'" *American Scientist* 57, no. 2 (1969): 244–68; Peter Fischer et al., "The Bystander-Effect: A Meta-Analytic Review on Bystander Intervention in Dangerous and Non-Dangerous Emergencies," *Psychological Bulletin* 137, no. 4 (2011): 517–37, https://doi.org/10.1037/a0023304.

20. Avner Ben-Ner and Amit Kramer, "Personality and Altruism in the Dictator Game: Relationship to Giving to Kin, Collaborators, Competitors, and Neutrals," *Personality and Individual Differences* 51, no. 3 (August 2011): 216–21, https://doi.org/10.1016/j.paid.2010.04.024; Ryo Oda et al., "Personality and Altruism in Daily Life," *Personality and Individual Differences* 56 (January 2014): 206–9, https://doi.org/10.1016/j.paid.2013.09.017; Dovidio et al., *The Social Psychology of Prosocial Behavior*; Fischer et al., "The Bystander-Effect."

21. Sarah Francis Smith et al., "Are Psychopaths and Heroes Twigs off the Same Branch? Evidence from College, Community, and Presidential Samples," *Journal of Research in Personality* 47, no. 5 (October 2013): 634–46, https://doi.org/10.1016/j.jrp.2013.05.006.

22. Daphna Oyserman, Heather M. Coon, and Markus Kemmelmeier, "Rethinking Individualism and Collectivism: Evaluation of Theoretical Assumptions and Meta-Analyses," *Psychological Bulletin* 128, no. 1 (2002): 3–72, https://doi.org/10.1037/0033-2909.128.1.3; Marilynn B. Brewer and Ya-Ru Chen, "Where (Who) Are Collectives in Collectivism? Toward Conceptual Clarification of Individualism and Collectivism," *Psychological Review* 114, no. 1 (2007): 133–51, https://doi.org/10.1037/0033-295X.114.1.133.

23. Mark Levine and Simon Crowther, "The Responsive Bystander: How Social Group Membership and Group Size Can Encourage as Well as Inhibit Bystander Intervention," *Journal of Personality and Social Psychology* 95, no. 6 (2008): 1429–39, https://doi.org/10.1037/a0012634; Fischer et al., "The Bystander-Effect."

24. Loren J. Martin et al., "Reducing Social Stress Elicits Emotional Contagion of Pain in Mouse and Human Strangers," *Current Biology* 25, no. 3 (February 2015): 326–32, https://doi.org/10.1016/j.cub.2014.11 .028.

25. Dovidio et al., *The Social Psychology of Prosocial Behavior*; Oda et al., "Personality and Altruism in Daily Life"; William John Ickes, ed., *Empathic Accuracy* (New York: Guilford Press, 1997); Bruce E. Chlopan et al., "Empathy: Review of Available Measures," *Journal of Personality and Social Psychology* 48, no. 3 (1985): 635–53, https://doi.org/10.1037 /0022-3514.48.3.635.

26. R. William Doherty, "The Emotional Contagion Scale: A Measure of Individual Differences," *Journal of Nonverbal Behavior* 21, no. 2 (1997): 131–54, https://doi.org/10.1023/A:1024956003661; Mark H. Davis, "Measuring Individual Differences in Empathy: Evidence for a Multidimensional Approach," *Journal of Personality and Social Psychology* 44, no. 1 (January 1983): 113–26, https://doi.org/10.1037/0022-3514.44.1.113; Louis A. Penner et al., "Measuring the Prosocial Personality," in *Advances in Personality Assessment* 10 (1995): 147–63.

27. Davis, "Measuring Individual Differences in Empathy."

28. C. D. Batson, "Altruism and Prosocial Behavior," in *The Handbook of Social Psychology*, ed. Daniel T. Gilbert, Susan T. Fiske, and Gardner Lindzey, 4th ed. (New York: Oxford University Press, 1998), 2:282–316.

29. Penner et al., "Measuring the Prosocial Personality."

30. Stephanie D. Preston et al., "Understanding Empathy and Its Disorders Through a Focus on the Neural Mechanism," *Cortex* 127 (2020): 347–70, https://doi.org/10.1016/j.cortex.2020.03.001.

31. Jennifer S. Beer, "Exaggerated Positivity in Self-Evaluation: A Social Neuroscience Approach to Reconciling the Role of Self-Esteem Protection and Cognitive Bias: Social Neuroscience of Exaggerated Positivity," *Social and Personality Psychology Compass* 8, no. 10 (October 2014): 583–94, https://doi.org/10.1111/spc3.12133.

32. Nancy Eisenberg and Randy Lennon, "Sex Differences in Empathy and Related Capacities," *Psychological Bulletin* 94, no. 1 (1983): 100–131.

33. Richard E. Nisbett and Timothy D. Wilson, "Telling More Than We Can Know: Verbal Reports on Mental Processes," *Psychological Review* 7 (1977): 231–59.

34. Ajzen, "The Theory of Planned Behavior."

35. C. Zahn-Waxler and M. Radke-Yarrow, "The Development of Altruism: Alternative Research Strategies," in *The Development of Prosocial Behavior*, ed. Nancy Eisenberg (New York: Academic Press, 1982), 133–62; Nancy Eisenberg and Richard A. Fabes, "Prosocial Development," in *Handbook of Child Psychology*, ed. Nancy Eisenberg, 5th ed. (New York: Wiley, 1998), 3:701–78; Nancy Eisenberg and Janet Strayer, *Empathy and Its Development* (Cambridge University Press, Cambridge, MA, 1990); Martin L. Hoffman, *Empathy and Moral Development: Implications for Caring and Justice* (New York: Cambridge University Press, 2000); Carolyn Zahn-Waxler et al., "Development of Concern for Others," *Developmental Psychology* 28, no. 1 (1992): 126–36; Carolyn Zahn-Waxler, Marian Radke-Yarrow, and Robert A. King, "Child Rearing and Children's Prosocial Initiations Toward Victims of Distress," *Child Development* 50, no. 2 (1979): 319–30.

36. Dario Maestripieri, "The Biology of Human Parenting: Insights from Nonhuman Primates," *Neuroscience & Biobehavioral Reviews* 23, no. 3 (1999): 411–22, https://doi.org/10.1016/S0149-7634(98)00042-6.

37. Marinus H. van IJzendoorn, "Attachment, Emergent Morality, and Aggression: Toward a Developmental Socioemotional Model of Antisocial Behaviour," *International Journal of Behavioral Development* 21, no. 4 (November 1997): 703–27, https://doi.org/10.1080/016502597384631.

38. R. J. R. Blair et al., "The Development of Psychopathy," *Journal of Child Psychology and Psychiatry* 47, nos. 3–4 (March 2006): 262–76, https://doi.org/10.1111/j.1469-7610.2006.01596.x.

39. Abigail A. Marsh and Robert James R. Blair, "Deficits in Facial Affect Recognition Among Antisocial Populations: A Meta-Analysis," *Neuroscience and Biobehavioral Reviews* 32 (2008): 454–65; R. J. R. Blair, "The Amygdala and Ventromedial Prefrontal Cortex: Functional Contributions and Dysfunction in Psychopathy," *Philosophical Transactions of the Royal Society of London. Series B: Biological Sciences* 363, no. 1503 (August 12, 2008): 2557–65, https://doi.org/10.1098/rstb.2008.0027; R. J. R. Blair, "The Amygdala and Ventromedial Prefrontal Cortex in Morality and Psychopathy," *Trends in Cognitive Sciences* 11, no. 9 (September 2007): 387–92, https://doi.org/10.1016/j.tics.2007.07.003.

40. Abigail A. Marsh, "Neural, Cognitive, and Evolutionary Foundations of Human Altruism," *Wiley Interdisciplinary Reviews: Cognitive Science* 7, no. 1 (2016): 59–71.

41. Peter Johansson and Margaret Kerr, "Psychopathy and Intelligence: A Second Look," *Journal of Personality Disorders* 19, no. 4 (August 2005): 357–69, https://doi.org/10.1521/pedi.2005.19.4.357.

42. Stephanie N. Mullins-Sweatt et al., "The Search for the Successful Psychopath," *Journal of Research in Personality* 44, no. 4 (August 2010): 554–58, https://doi.org/10.1016/j.jrp.2010.05.010.

43. Preston et al., "Understanding Empathy and Its Disorders."

44. Lai Ling Chau et al., "Intrinsic and Extrinsic Religiosity as Related to Conscience, Adjustment, and Altruism," *Personality and Individual Differences* 11, no. 4 (1990): 397–400, https://doi.org/10.1016/0191 -8869(90)90222-D; H. Lovell Smith, Anthony Fabricatore, and Mark Peyrot, "Religiosity and Altruism Among African American Males: The Catholic Experience," *Journal of Black Studies* 29, no. 4 (March 1999): 579–97, https://doi.org/10.1177/002193479902900407.

45. Chau et al., "Intrinsic and Extrinsic Religiosity."

8. COMPARING THE ALTRUISTIC RESPONSE MODEL TO OTHER THEORIES

1. W. D. Hamilton, "The Evolution of Altruistic Behavior." *The American Naturalist* 97, no. 896 (1963): 354–56.

2. J. Maynard Smith, "Group Selection and Kin Selection," *Nature* 201 (1964): 1145–47.

3. B. R. Vickers et al., "Motor System Engagement in Charitable Giving: The Offspring Care Effect," forthcoming.

4. David S. Wilson, "A Theory of Group Selection," *Proceedings of the National Academy of Sciences USA* 72, no. 1 (January 1975): 143–46; Eugene Burnstein, Christian Crandall, and Shinobu Kitayama, "Some Neo-Darwinian Decision Rules for Altruism: Weighing Cues for Inclusive Fitness as a Function of the Biological Importance of the Decision," *Journal of Personality and Social Psychology* 67, no. 5 (1994): 773–89, https:// doi.org/10.1037/0022-3514.67.5.773.

5. Jonathan Birch, "Are Kin and Group Selection Rivals or Friends?" *Current Biology* 29, no. 11 (June 2019): R433–38, https://doi.org/10 .1016/j.cub.2019.01.065; David S. Wilson and Lee A. Dugatkin, "Group Selection and Assortative Interactions," *The American Naturalist* 149, no. 2 (February 1, 1997): 336–51, https://doi.org/10.1086/285993.

6. Stephanie D. Preston and Frans B. M. de Waal, "Altruism," in *The Handbook of Social Neuroscience*, ed. Jean Decety and John T. Cacioppo (New York: Oxford University Press, 2011), 565–85.

7. Peter Fischer et al., "The Unresponsive Bystander: Are Bystanders More Responsive in Dangerous Emergencies?" *European Journal of Social Psychology* 36, no. 2 (March 2006): 267–78, https://doi.org/10.1002/ejsp.297; Bibb Latané and John M. Darley, "Bystander 'Apathy.'" *American Scientist* 57, no. 2 (1969): 244–68.

8. Herbert Gintis, "Strong Reciprocity and Human Sociality," *Journal of Theoretical Biology* 206, no. 2 (2000): 169–79, https://doi.org/10.1006/jtbi .2000.2111; Samuel Bowles and Herbert Gintis, "The Evolution of Strong Reciprocity: Cooperation in Heterogeneous Populations," *Theoretical Population Biology* 65, no. 1 (2004): 17–28, https://doi.org/10.1016 /j.tpb.2003.07.001; Ernst Fehr, Urs Fischbacher, and Simon Gächter, "Strong Reciprocity, Human Cooperation, and the Enforcement of Social Norms," *Human Nature* 13, no. 1 (2002): 1–25, https://doi.org/10 .1007/s12110-002-1012-7.

9. Philip G. Zimbardo, "On 'Obedience to Authority,'" *American Psychologist* 29, no. 7 (1974): 567, https://doi.org/10.1037/h0038158.

10. I'm sure you could link it to a gene or set of genes, but that does not mean the gene encodes for cooperation in humans, rather than some more generic feature like learning and working memory.

11. Stanley Milgram, *Obedience to Authority: An Experimental View* (New York: Harper & Row, 1974).

12. Philip G. Zimbardo, Christina Maslach, and Craig Haney, "Reflections on the Stanford Prison Experiment: Genesis, Transformations, Consequences," in *Obedience to Authority: Current Perspectives on the Milgram Paradigm*, ed. T. Blass (Hoboken, NJ: Erlbaum, 1999), 193–237.

13. It should be noted that some people question the veracity of Zimbardo's claims and suggest that the guards were either instructed to act cruelly or did not exhibit such alarming behavior.

14. Frans B. M. de Waal, *Peacemaking Among Primates* (Cambridge, MA: Harvard University Press, 1989).

15. Peter Verbeek and Frans B. M. de Waal, "Peacemaking Among Preschool Children," *Peace and Conflict: Journal of Peace Psychology* 7, no. 1 (2001): 5–28, https://doi.org/10.1207/S15327949PAC0701_02.

16. Latané and Darley, "Bystander 'Apathy.'"

17. Martin Gansberg, "Thirty-Eight Who Saw Murder Didn't Call the Police," *New York Times* 27 (1964).

18. Samuel P. Oliner, "Extraordinary Acts of Ordinary People," in *Altruism and Altruistic Love: Science, Philosophy, and Religion in Dialogue*, ed. Steven Post et al. (Oxford: Oxford University Press, 2002), 123–39.

19. Rachel Manning, Mark Levine, and Alan Collins, "The Kitty Genovese Murder and the Social Psychology of Helping: The Parable of the 38 Witnesses," *American Psychologist* 62, no. 6 (2007): 555.

20. C. D. Batson, *The Altruism Question: Toward A Social-Psychological Answer* (New York: Taylor & Francis, 2014); C. D. Batson, *Altruism in Humans* (New York: Oxford University Press, 2011).

21. Carolyn Zahn-Waxler and Marian Radke-Yarrow, "The Development of Altruism: Alternative Research Strategies," in *The Development of Prosocial Behavior*, ed. Nancy Eisenberg (New York: Academic Press, 1982), 133–62.

22. Carolyn Zahn-Waxler, Barbara Hollenbeck, and Marian Radke-Yarrow, "The Origins of Empathy and Altruism," in *Advances in Animal Welfare Science*, ed. Michael W. Fox and Linda D. Mickley (Washington, DC: Humane Society of the United States, 1984), 21–39.

23. Frans B. M. de Waal and Stephanie D. Preston, "Mammalian Empathy: Behavioural Manifestations and Neural Basis," *Nature Reviews Neuroscience* 18, no. 8 (2017): 498–510.

24. I. Morrison et al., "Vicarious Responses to Pain in Anterior Cingulate Cortex: Is Empathy a Multisensory Issue?" *Cognitive, Affective, and Behavioral Neuroscience* 4, no. 2 (June 2004): 270–78; T. Singer et al., "Empathy for Pain Involves the Affective but Not Sensory Components of Pain," *Science* 303, no. 5661 (February 20, 2004): 1157–62.

25. Claus Lamm, Jean Decety, and Tania Singer, "Meta-Analytic Evidence for Common and Distinct Neural Networks Associated with Directly Experienced Pain and Empathy for Pain," *Neuroimage* 54, no. 3 (2011): 2492–502.

26. Pascal Molenberghs, "The Neuroscience of In-Group Bias," *Neuroscience & Biobehavioral Reviews* 37, no. 8 (September 2013): 1530–36, https://doi.org/10.1016/j.neubiorev.2013.06.002; Yawei Cheng et al., "Expertise Modulates the Perception of Pain in Others," *Current Biology* 17, no. 19 (October 9, 2007): 1708–13, https://doi.org/10.1016/j.cub.2007.09.020.

27. Stephanie D. Preston et al., "The Neural Substrates of Cognitive Empathy," *Social Neuroscience* 2, nos. 3–4 (2007): 254–75, https://doi.org/10.1080 /17470910701376902.

28. Stephanie D. Preston and Frans B. M. de Waal, "Empathy: Its Ultimate and Proximate Bases," *Behavioral and Brain Sciences* 25, no. 1 (2002): 1–71, https://doi.org/10.1017/S0140525X02000018.

29. Paul Bloom, *Against Empathy: The Case for Rational Compassion* (New York: Ecco, 2016).

30. Tony W. Buchanan and Stephanie D. Preston, "Stress Leads to Prosocial Action in Immediate Need Situations," *Frontiers in Behavioral Neuroscience* 8, no. 5 (2014), https://doi.org/10.3389/fnbeh.2014.00005.

31. Norbert Schwarz and Gerald L. Clore, "Mood as Information: 20 Years Later," *Psychological Inquiry* 14, no. 3–4 (2003): 296–303; Antonio Damasio, *Descartes' Error : Emotion, Reason, and the Human Brain* (New York: Putnam, 1994); Paul Slovic and Ellen Peters, "Risk Perception and Affect," *Current Directions in Psychological Science* 15, no. 6 (December 2006): 322–25, https://doi.org/10.1111/j.1467-8721.2006.00461.x; Jennifer S. Lerner et al., "Emotion and Decision Making," *Annual Review of Psychology* 66, no. 1 (January 3, 2015): 799–823, https://doi.org/10.1146 /annurev-psych-010213-115043; G. F. Loewenstein et al., "Risk as Feelings," *Psychological Bulletin* 127, no. 2 (2001): 267–86, https://doi.org/10 .1037/0033-2909.127.2.267; Filippo Aureli and Colleen M. Schaffner, "Relationship Assessment Through Emotional Mediation," *Behaviour* 139, nos. 2–3 (2002): 393–420. There are multiple named theories like this, including Antonio Damasio's somatic marker hypothesis, George Lowenstein and colleagues' risk-as-feelings model, Jennifer Lerner's emotion-imbued choice model, Norbert Schwartz and Gerald Clore's mood as information model, Paul Slovic and Ellen Peter's affect heuristic, and Filippo Aureli's emotionally mediated helping.

CONCLUSION

1. William E. Wilsoncroft, "Babies by Bar-Press: Maternal Behavior in the Rat," *Behavior Research Methods, Instruments and Computers* 1 (1969): 229–30; Michael Numan, "Motivational Systems and the Neural Circuitry of Maternal Behavior in the Rat," *Developmental Psychobiology* 49, no. 1 (January 2007): 12–21.

2. J. S. Rosenblatt, "Nonhormonal Basis of Maternal Behavior in the Rat," *Science* 156 (1967): 1512–14; J. S. Rosenblatt and K. Ceus, "Estrogen Implants in the Medial Preoptic Area Stimulate Maternal Behavior in Male Rats," *Hormones and Behavior* 33 (1998): 23–30.

3. Selwyn W. Becker and Alice H. Eagly, "The Heroism of Women and Men," *American Psychologist* 59, no. 3 (2004): 163–78, https://doi.org/10.1037/0003-066x.59.3.163; Leonardo Christov-Moore et al., "Empathy: Gender Effects in Brain and Behavior," *Neuroscience & Biobehavioral Reviews* 46, Part 4 (2014): 604–27, https://doi.org/10.1016/j.neubiorev.2014.09.001; Mark Coultan, "NY Toasts Subway Superman After Death-Defying Rescue," *The Age*, January 6, 2007, http://www.theage.com.au/news/world/ny-toasts-subway-superman-after-deathdefying-rescue/2007/01/05/1167777281613.html.

4. Lyn Craig, "Does Father Care Mean Fathers Share? A Comparison of How Mothers and Fathers in Intact Families Spend Time with Children," *Gender & Society* 20, no. 2 (April 2006): 259–81, https://doi.org/10.1177/0891243205285212.

5. Heather E. Ross et al., "Variation in Oxytocin Receptor Density in the Nucleus Accumbens Has Differential Effects on Affiliative Behaviors in Monogamous and Polygamous Voles," *The Journal of Neuroscience* 29, no. 5 (February 4, 2009): 1312–18, https://doi.org/10.1523/JNEUROSCI.5039-08.2009.

6. Michael Numan and Thomas R. Insel, *The Neurobiology of Parental Behavior* (New York: Springer, 2003).

7. Jules B. Panksepp and Garet P. Lahvis, "Rodent Empathy and Affective Neuroscience," *Neuroscience & Biobehavioral Reviews* 35, no. 9 (October 2011): 1864–75, https://doi.org/10.1016/j.neubiorev.2011.05.013.

8. Pascal Molenberghs, "The Neuroscience of In-Group Bias," *Neuroscience & Biobehavioral Reviews* 37, no. 8 (September 2013): 1530–36, https://doi.org/10.1016/j.neubiorev.2013.06.002.

9. Mark J. Pletcher et al., "Trends in Opioid Prescribing by Race/Ethnicity for Patients Seeking Care in US Emergency Departments," *Journal of the American Medical Association* 299, no. 1 (January 2, 2008): 70–78, https://doi.org/10.1001/jama.2007.64; Brian B. Drwecki et al., "Reducing Racial Disparities in Pain Treatment: The Role of Empathy and Perspective-Taking," *Pain* 152, no. 5 (May 1, 2011): 1001–6, https://doi.org/10.1016/j.pain.2010.12.005; Kelly M. Hoffman et al.,

"Racial Bias in Pain Assessment and Treatment Recommendations, and False Beliefs About Biological Differences Between Blacks and Whites," *Proceedings of the National Academy of Sciences of the United States of America* 113, no. 16 (April 19, 2016): 4296–4301, https://doi.org/10.1073/pnas.1516047113.

10. Stephanie D. Preston and Frans B. M. de Waal, "Empathy: Its Ultimate and Proximate Bases," *Behavioral and Brain Sciences* 25, no. 1 (2002): 1–71.

11. Robert L. Trivers, "The Evolution of Reciprocal Altruism," *Quarterly Review of Biology* 46 (1971): 35–57.

12. Irene Pepperberg, *The Alex Studies: Cognitive and Communicative Abilities of Grey Parrots* (Cambridge, MA: Harvard University Press, 2009).

13. Tony W. Buchanan et al., "The Empathic, Physiological Resonance of Stress," *Social Neuroscience* 7, no. 2 (2012): 191–201, https://doi.org/10.1080/17470919.2011.588723; Emilie C. Perez et al., "Physiological Resonance Between Mates Through Calls as Possible Evidence of Empathic Processes in Songbirds," *Hormones and Behavior* 75 (September 1, 2015): 130–41, https://doi.org/10.1016/j.yhbeh.2015.09.002; Joanne L. Edgar and Christine J. Nicol, "Socially-Mediated Arousal and Contagion Within Domestic Chick Broods," *Scientific Reports* 8, no. 1 (December 2018): 10509, https://doi.org/10.1038/s41598-018-28923-8.

14. Katherine Taylor et al., "Precision Rescue Behavior in North American Ants," *Evolutionary Psychology* 11, no. 3 (July 2013): 14747049130110012.

15. Seweryn Olkowicz et al., "Birds Have Primate-like Numbers of Neurons in the Forebrain," *Proceedings of the National Academy of Sciences* 113, no. 26 (June 28, 2016): 7255–60, https://doi.org/10.1073/pnas.1517131113.

16. Sandeep Gupta et al., "Defining Structural Homology between the Mammalian and Avian Hippocampus through Conserved Gene Expression Patterns Observed in the Chick Embryo," *Developmental Biology* 366, no. 2 (June 15, 2012): 125–41, https://doi.org/10.1016/j.ydbio.2012.03.027; Olkowicz et al., "Birds Have Primate-like Numbers of Neurons in the Forebrain."

17. Kristina Simonyan, Barry Horwitz, and Erich D. Jarvis, "Dopamine Regulation of Human Speech and Bird Song: A Critical Review," *Brain*

and Language 122, no. 3 (September 1, 2012): 142–50, https://doi.org/10 .1016/j.bandl.2011.12.009.

18. David Kabelik and D. Sumner Magruder, "Involvement of Different Mesotocin (Oxytocin Homologue) Populations in Sexual and Aggressive Behaviours of the Brown Anole," *Biology Letters* 10, no. 8 (August 31, 2014): 20140566, https://doi.org/10.1098/rsbl.2014.0566.

19. James L. Goodson, Aubrey M. Kelly, and Marcy A. Kingsbury, "Evolving Nonapeptide Mechanisms of Gregariousness and Social Diversity in Birds," *Hormones and Behavior* 61, no. 3 (March 2012): 239–50, https://doi.org/10.1016/j.yhbeh.2012.01.005.

20. J. D. Reynolds, N. B. Goodwin, and R. P. Freckleton, "Evolutionary Transitions in Parental Care and Live Bearing in Vertebrates," ed. S. Balshine, B. Kempenaers, and T. Székely, *Philosophical Transactions of the Royal Society of London. Series B: Biological Sciences* 357, no. 1419 (March 29, 2002): 269–81, https://doi.org/10.1098/rstb.2001.0930.

21. Reynolds, Goodwin, and Freckleton, "Evolutionary Transitions."

22. Tim H. Clutton-Brock, *The Evolution of Parental Care* (Princeton, NJ: Princeton University Press, 1991).

23. Stephen E. G. Lea and Paul Webley, "Money as Tool, Money as Drug: The Biological Psychology of a Strong Incentive," *Behavioral and Brain Sciences* 29, no. 02 (2006): 161–209.

REFERENCES

Acebo, Christine, and Evelyn B. Thoman. "Role of Infant Crying in the Early Mother-Infant Dialogue." *Physiology & Behavior* 57, no. 3 (1995): 541–47.

Addessi, Elsa, Amy T. Galloway, Elisabetta Visalberghi, and Leann L. Birch. "Specific Social Influences on the Acceptance of Novel Foods in 2–5-Year-Old Children." *Appetite* 45, no. 3 (December 2005): 264–71. https://doi.org/10.1016/j.appet.2005.07.007.

Ajzen, Icek. "The Theory of Planned Behavior." *Organizational Behavior and Human Decision Processes* 50, no. 2 (December 1991): 179–211. https://doi .org/10.1016/0749-5978(91)90020-T.

Andreoni, James. "Impure Altruism and Donations to Public Goods: A Theory of Warm-Glow Giving." *The Economic Journal* 100, no. 401 (1990): 464–77.

Andreoni, James, William T. Harbaugh, and Lise Vesterlund. "Altruism in Experiments." In *The New Palgrave Dictionary of Economics*, ed. Steven N. Durlauf and Lawrence E. Bloom, 134–38. London: Palgrave Macmillan, 2008.

Anik, Lalin, Lara B. Aknin, Michael I. Norton, and Elizabeth W. Dunn. "Feeling Good About Giving: The Benefits (and Costs) of Self-Interested Charitable Behavior." *SSRN Electronic Journal* 2009. https://doi.org/10 .2139/ssrn.1444831.

Associated Press. "Dog Dies After Saving Trinidad Man from Fire." *Los Angeles Times*, October 11, 2008.

Aureli, Filippo, Stephanie D. Preston, and Frans B. M. de Waal. "Heart Rate Responses to Social Interactions in Free-Moving Rhesus Macaques (Macaca Mulatta): A Pilot Study." *Journal of Comparative Psychology* 113, no. 1 (March 1999): 59–65.

Aureli, Filippo, and Colleen M. Schaffner. "Relationship Assessment Through Emotional Mediation." *Behaviour* 139, nos. 2–3 (2002): 393–420.

Azar, Beth. "Nature, Nurture: Not Mutually Exclusive." *APA Monitor* 28 (1997): 1–28.

Bakermans-Kranenburg, Marian J., and Marinus H. van IJzendoorn. "A Sociability Gene? Meta-Analysis of Oxytocin Receptor Genotype Effects in Humans." *Psychiatric Genetics* 24, no. 2 (April 2014): 45–51. https://doi .org/10.1097/YPG.0b013e3283643684.

Bandura, Albert. "Self-Efficacy." In *The Corsini Encyclopedia of Psychology*, ed. Irving B. Weiner and W. Edward Craighead, 1–3. Hoboken, NJ: Wiley, 2010. https://doi.org/10.1002/9780470479216.corpsy0836.

Barclay, Pat. "Altruism as a Courtship Display: Some Effects of Third-Party Generosity on Audience Perceptions." *British Journal of Psychology* 101, no. 1 (2010): 123–35. https://doi.org/10.1348/000712609X435733.

Baron-Cohen, Simon, and Sally Wheelwright. "The Empathy Quotient: An Investigation of Adults with Asperger Syndrome or High Functioning Autism, and Normal Sex Differences." *Journal of Autism and Developmental Disorders* 34, no. 2 (April 2004): 163–75. https://doi.org/10.1023/B:JADD .0000022607.19833.00.

Batson, C. D. "Altruism and Prosocial Behavior." In *The Handbook of Social Psychology*, ed. Daniel T. Gilbert, Susan T. Fiske, and Gardner Lindzey, 4th ed., 2:282–316. New York: Oxford University Press, 1998.

——. *Altruism in Humans*. New York: Oxford University Press, 2011.

——. *The Altruism Question: Toward a Social-Psychological Answer*. New York: Taylor & Francis, 2014.

Batson, C. Daniel. "The Naked Emperor: Seeking a More Plausible Genetic Basis for Psychological Altruism." *Economics and Philosophy* 26, no. 2 (2010): 149–64. https://doi.org/10.1017/S0266267110000179.

Batson, C. Daniel, and Jay S. Coke. "Empathy: A Source of Altruistic Motivation for Helping." In *Altruism and Helping Behavior*, ed. J. Philippe Rushton and Richard M. Sorrentino, 167–87.

Baumgartner, Thomas, Markus Heinrichs, Aline Vonlanthen, Urs Fischbacher, and Ernst Fehr. "Oxytocin Shapes the Neural Circuitry of Trust and Trust Adaptation in Humans." *Neuron* 58, no. 4 (May 22, 2008): 639–50.

Bechara, Antonio, Hanna Damasio, Daniel Tranel, and Steven Anderson. "Dissociation of Working Memory from Decision Making Within the Human Prefrontal Cortex." *Journal of Neuroscience* 18 (1998): 428–37.

Bechara, Antoine, Hanna Damasio, and Antonio R. Damasio. "Emotion, Decision Making and the Orbitofrontal Cortex." *Cerebral Cortex* 10, no. 3 (2000): 295–307.

Becker, Jill B., and Jane R. Taylor. "Sex Differences in Motivation." In *Sex Differences in the Brain: From Genes to Behavior*, ed. J. B. Becker, K. J. Berkley, N. Geary, E. Hampson, J. P. Herman, and E. A. Young, 177–99. New York: Oxford University Press, 2008.

Becker, Selwyn W., and Alice H. Eagly. "The Heroism of Women and Men." *American Psychologist* 59, no. 3 (2004): 163–78. https://doi.org/10.1037/0003 -066x.59.3.163.

Beeman, William O. "Making Grown Men Weep." In *Aesthetics in Performance: Formations of Symbolic Instruction and Experience*, ed. Angela Hobart and Bruce Kapferer, 23–42. New York: Berghahn Books, 2005.

Beer, Jennifer S. "Exaggerated Positivity in Self-Evaluation: A Social Neuroscience Approach to Reconciling the Role of Self-Esteem Protection and Cognitive Bias—Social Neuroscience of Exaggerated Positivity." *Social and Personality Psychology Compass* 8, no. 10 (October 2014): 583–94. https://doi.org/10.1111/spc3.12133.

Ben-Ner, Avner, and Amit Kramer. "Personality and Altruism in the Dictator Game: Relationship to Giving to Kin, Collaborators, Competitors, and Neutrals." *Personality and Individual Differences* 51, no. 3 (August 2011): 216–21. https://doi.org/10.1016/j.paid.2010.04.024.

Berridge, Kent C., and Terry E. Robinson. "What Is the Role of Dopamine in Reward: Hedonic Impact, Reward Learning, or Incentive Salience?" *Brain Research Reviews* 28, no. 3 (December 1998): 309–69.

Berry, Diane S., and Leslie Z. McArthur. "Some Components and Consequences of a Babyface." *Journal of Personality and Social Psychology* 48, no. 2 (1985): 312–23. https://doi.org/10.1037/0022-3514.48.2.312.

Birch, Jonathan. "Are Kin and Group Selection Rivals or Friends?" *Current Biology* 29, no. 11 (June 2019): R433–38. https://doi.org/10.1016/j.cub.2019 .01.065.

Blair, R. James R. "The Amygdala and Ventromedial Prefrontal Cortex: Functional Contributions and Dysfunction in Psychopathy." *Philosophical Transactions of the Royal Society of London. Series B: Biological Sciences* 363, no. 1503 (August 12, 2008): 2557–65. https://doi.org/10.1098 /rstb.2008.0027.

———. "The Amygdala and Ventromedial Prefrontal Cortex in Morality and Psychopathy." *Trends in Cognitive Science* 11, no. 9 (September 2007): 387–92. https://doi.org/10.1016/j.tics.2007.07.003.

Blair, R. James R., Karina S. Peschardt, Salima Budhani, Derek G. V. Mitchell, and Daniel S. Pine. "The Development of Psychopathy." *Journal of Child Psychology and Psychiatry* 47, nos. 3–4 (March 2006): 262–76. https://doi.org/10.1111/j.1469-7610.2006.01596.x.

Bloom, Paul. *Against Empathy: The Case for Rational Compassion*. New York: Ecco, 2016.

Bowlby, John. *Attachment and Loss*, Volume 1: *Attachment*. New York: Basic Books, 1969.

———. "The Nature of the Child's Tie to His Mother." *International Journal of Psycho-Analysis* 39 (1958): 350–73.

Bowles, Samuel, and Herbert Gintis. "The Evolution of Strong Reciprocity: Cooperation in Heterogeneous Populations." *Theoretical Population Biology* 65, no. 1 (2004): 17–28. https://doi.org/10.1016/j.tpb.2003.07.001.

Brembs, Björn, and Jan Wiener. "Context and Occasion Setting in Drosophila Visual Learning." *Learning & Memory* 13, no. 5 (September 1, 2006): 618–28. https://doi.org/10.1101/lm.318606.

Brennan, Kelly A., and Phillip R. Shaver. "Dimensions of Adult Attachment, Affect Regulation, and Romantic Relationship Functioning." *Personality and Social Psychology Bulletin* 21, no. 3 (1995): 267–83.

Brewer, Marilynn B., and Ya-Ru Chen. "Where (Who) Are Collectives in Collectivism? Toward Conceptual Clarification of Individualism and Collectivism." *Psychological Review* 114, no. 1 (2007): 133–51. https://doi.org/10.1037/0033-295X.114.1.133.

Broad, Kevin D., James P. Curley, and Eric B. Keverne. "Mother-Infant Bonding and the Evolution of Mammalian Social Relationships." *Philosophical Transactions of the Royal Society of London. Series B: Biological Sciences* 361, no. 1476 (December 29, 2006): 2199–214. https://doi.org/10.1098/rstb.2006.1940.

Brown, Jennifer R., Hong Ye, Roderick T. Bronson, Pieter Dikkes, and Michael E. Greenberg. "A Defect in Nurturing in Mice Lacking the Immediate Early Gene FosB." *Cell* 86, no. 2 (1996): 297–309.

Brown, Stephanie L., and R. Michael Brown. "Connecting Prosocial Behavior to Improved Physical Health: Contributions from the Neurobiology of Parenting." *Neuroscience & Biobehavioral Reviews* 55 (August 2015): 1–17. https://doi.org/10.1016/j.neubiorev.2015.04.004.

Brown, Stephanie L., R. Michael Brown, and Louis A. Penner. *Moving Beyond Self-Interest: Perspectives from Evolutionary Biology, Neuroscience, and the Social Sciences.* New York: Oxford University Press, 2011.

Buchanan, Tony W., Sara L. Bagley, R. Brent Stansfield, and Stephanie D. Preston. "The Empathic, Physiological Resonance of Stress." *Social Neuroscience* 7, no. 2 (2012): 191–201. https://doi.org/10.1080/17470919.2011.588723.

Buchanan, Tony W., and Stephanie D. Preston. "Stress Leads to Prosocial Action in Immediate Need Situations." *Frontiers in Behavioral Neuroscience* 8, no. 5 (2014). https://doi.org/10.3389/fnbeh.2014.00005.

Buckley, Cara. "Man Is Rescued by Stranger on Subway Tracks." *New York Times*, January 3, 2007.

Burkett, James P., Elissar Andari, Zachary V. Johnson, Daniel C. Curry, Frans B .M. de Waal, and Larry J. Young. "Oxytocin-Dependent Consolation Behavior in Rodents." *Science* 351, no. 6271 (2016): 375–78.

Burnstein, Eugene, Christian Crandall, and Shinobu Kitayama. "Some Neo-Darwinian Decision Rules for Altruism: Weighing Cues for Inclusive Fitness as a Function of the Biological Importance of the Decision." *Journal of Personality and Social Psychology* 67, no. 5 (1994): 773–89. https://doi.org/10.1037/0022-3514.67.5.773.

Carroll, Sean B. "Evo-Devo and an Expanding Evolutionary Synthesis: A Genetic Theory of Morphological Evolution." *Cell* 134, no. 1 (July 2008): 25–36. https://doi.org/10.1016/j.cell.2008.06.030.

Carstensen, Laura L., John M. Gottman, and Robert W. Levenson. "Emotional Behavior in Long-Term Marriage." *Psychology and Aging* 10, no. 1 (1995): 140–49. https://doi.org/10.1037/0882-7974.10.1.140.

Champagne, Frances A. "Epigenetic Mechanisms and the Transgenerational Effects of Maternal Care." *Frontiers in Neuroendocrinology* 29 (2008): 386–97.

Champagne, Frances A., Darlene D. Francis, Adam Mar, and Michael J. Meaney. "Variations in Maternal Care in the Rat as a Mediating Influence for the Effects of Environment on Development." *Physiology & Behavior* 79, no. 3 (2003): 359–71.

Chau, Lai Ling, Ronald C. Johnson, John K. Bowers, Thomas J. Darvill, and George P. Danko. "Intrinsic and Extrinsic Religiosity as Related to

Conscience, Adjustment, and Altruism." *Personality and Individual Differences* 11, no. 4 (1990): 397–400. https://doi.org/10.1016/0191-8869(90)90222-D.

Cheng, Yawei, Ching-Po Lin, Ho-Ling Liu, Yuan-Yu Hsu, Kun-Eng Lim, Daisy Hung, and Jean Decety. "Expertise Modulates the Perception of Pain in Others." *Current Biology* 17, no. 19 (October 9, 2007): 1708–13. https://doi.org/10.1016/j.cub.2007.09.020.

Chlopan, Bruce E., Marianne L. McCain, Joyce L. Carbonell, and Richard L. Hagen. "Empathy: Review of Available Measures." *Journal of Personality and Social Psychology* 48, no. 3 (1985): 635–53. https://doi.org/10.1037/0022-3514.48.3.635.

Christov-Moore, Leonardo, Elizabeth A. Simpson, Gino Coudé, Kristina Grigaityte, Marco Iacoboni, and Pier Francesco Ferrari. "Empathy: Gender Effects in Brain and Behavior," *Neuroscience & Biobehavioral Reviews* 46, Part 4 (2014): 604–27. https://doi.org/10.1016/j.neubiorev.2014.09.001.

Clayton, Nicola S., Juan C. Reboreda, and Alex Kacelnik. "Seasonal Changes of Hippocampus Volume in Parasitic Cowbirds." *Behavioural Processes* 41, no. 3 (December 1997): 237–43. https://doi.org/10.1016/S0376-6357(97)00050-8.

Clutton-Brock, Tim H. *The Evolution of Parental Care*. Princeton, NJ: Princeton University Press, 1991.

Connell, Sharon, John Fien, Jenny Lee, Helen Sykes, and David Yencken. " 'If It Doesn't Directly Affect You, You Don't Think About It': A Qualitative Study of Young People's Environmental Attitudes in Two Australian Cities." *Environmental Education Research* 5, no. 1 (February 1999): 95–113. https://doi.org/10.1080/1350462990050106.

Coultan, Mark. "NY Toasts Subway Superman After Death-Defying Rescue." *The Age*, January 6, 2007. http://www.theage.com.au/news/world/ny-toasts-subway-superman-after-deathdefying-rescue/2007/01/05/1167777281613.html.

Craig, Lyn. "Does Father Care Mean Fathers Share? A Comparison of How Mothers and Fathers in Intact Families Spend Time with Children." *Gender & Society* 20, no. 2 (April 2006): 259–81. https://doi.org/10.1177/0891243205285212.

Crocker, Jennifer, Amy Canevello, and Ashley A. Brown. "Social Motivation: Costs and Benefits of Selfishness and Otherishness." *Annual Review of Psychology* 68, no. 1 (January 3, 2017): 299–325. https://doi.org/10.1146 /annurev-psych-010416-044145.

Curtiss, Susan, and Harry A. Whitaker. *Genie: A Psycholinguistic Study of a Modern-Day Wild Child.* St. Louis, MO: Elsevier Science, 2014.

Damasio, Antonio. *Descartes' Error : Emotion, Reason, and the Human Brain.* New York: Putnam, 1994.

Darley, John M., and Bibb Latané. "Bystander Intervention in Emergencies: Diffusion of Responsibility." *Journal of Personality and Social Psychology* 8, no. 4 (1968): 377–83.

Darwin, Charles. *The Expression of the Emotions in Man and Animals.* 3rd ed. Oxford: Oxford University Press, 1872.

Davis, Mark H. "Measuring Individual Differences in Empathy: Evidence for a Multidimensional Approach." *Journal of Personality and Social Psychology* 44, no. 1 (January 1983): 113–26. https://doi.org/10.1037/0022-3514 .44.1.113.

Dawkins, Richard. *The Selfish Gene.* Oxford: Oxford University Press, 1976.

de Quervain, Dominique JF, Urs Fischbacher, Valerie Treyer, Melanie Schellhammer, Ulrich Schnyder, Alfred Buck, and Ernst Fehr. "The Neural Basis of Altruistic Punishment." *Science* 305, no. 5688 (August 27, 2004): 1254–58.

DeSteno, David, Monica Y. Bartlett, Jolie Baumann, Lisa A. Williams, and Leah Dickens. "Gratitude as Moral Sentiment: Emotion-Guided Cooperation in Economic Exchange." *Emotion* 10, no. 2 (2010): 289–93. https:// doi.org/10.1037/a0017883.

de Waal, Frans B. M. *Good Natured: The Origins of Right and Wrong in Humans and Other Animals.* Cambridge, MA: Harvard University Press, 1996.

——. *Peacemaking Among Primates.* Cambridge, MA: Harvard University Press, 1989.

——. "Putting the Altruism Back Into Altruism: The Evolution of Empathy." *Annual Review of Psychology* 59 (2008): 279–300.

de Waal, Frans B. M., and Filippo Aureli. "Consolation, Reconciliation, and a Possible Cognitive Difference Between Macaque and Chimpanzee." In *Reaching Into Thought: The Minds of the Great Apes,* ed. K. A. Bard, A. E. Russon, and S. T. Parker, 80–110. Cambridge: Cambridge University Press, 1996.

de Waal, Frans B. M., and Stephanie D. Preston. "Mammalian Empathy: Behavioural Manifestations and Neural Basis." *Nature Reviews Neuroscience* 18, no. 8 (2017): 498–510.

Dickert, Stephan, Daniel Västfjäll, Janet Kleber, and Paul Slovic. "Scope Insensitivity: The Limits of Intuitive Valuation of Human Lives in Public Policy." *Journal of Applied Research in Memory and Cognition* 4, no. 3 (2015): 248–55.

Doherty, R. William. "The Emotional Contagion Scale: A Measure of Individual Differences." *Journal of Nonverbal Behavior* 21, no. 2 (1997): 131–54. https://doi.org/10.1023/A:1024956003661.

Dovidio, John F., Jane Allyn Piliavin, David A. Schroeder, and Louis A. Penner. *The Social Psychology of Prosocial Behavior.* Philadelphia: Erlbaum, 2006.

Drwecki, Brian B., Colleen F. Moore, Sandra E. Ward, and Kenneth M. Prkachin. "Reducing Racial Disparities in Pain Treatment: The Role of Empathy and Perspective-Taking." *Pain* 152, no. 5 (May 1, 2011): 1001–6. https://doi.org/10.1016/j.pain.2010.12.005.

Dunfield, Kristen A. "A Construct Divided: Prosocial Behavior as Helping, Sharing, and Comforting Subtypes." *Frontiers in Psychology* 5 (September 2, 2014): 958. https://doi.org/10.3389/fpsyg.2014.00958.

Dunn, Elizabeth W., Laura B. Aknin, and Michael I. Norton. "Spending Money on Others Promotes Happiness." *Science* 319, no. 5870 (March 21, 2008): 1687–88. https://doi.org/10.1126/science.1150952.

Dyer, Carmel Bitondo, Valory N. Pavlik, Kathleen Pace Murphy, and David J. Hyman. "The High Prevalence of Depression and Dementia in Elder Abuse or Neglect." *Journal of the American Geriatrics Society* 48, no. 2 (February 2000): 205–8. https://doi.org/10.1111/j.1532-5415.2000.tb03913.x.

Edgar, Joanne L., and Christine J. Nicol. "Socially-Mediated Arousal and Contagion Within Domestic Chick Broods." *Scientific Reports* 8, no. 1 (December 2018): 10509. https://doi.org/10.1038/s41598-018-28923-8.

Eibl-Eibesfeldt, Irenaus. *Love and Hate.* Trans. Geoffrey Strachan. 2nd ed. New York: Schocken Books, 1971.

Einon, Dorothy, and Michael Potegal. "Temper Tantrums in Young Children." In *The Dynamics of Aggression: Biological and Social Processes in Dyads and Groups,* ed. Michael Potegal and John F. Knutson, 157–94. New York: Psychology Press, 1994.

Eisenberg, Nancy, and Richard A. Fabes. "Prosocial Development." In *Handbook of Child Psychology,* ed. Nancy Eisenberg, 5th ed., 3:701–78. New York: Wiley, 1998.

Eisenberg, Nancy, Richard A. Fabes, Bridget Murphy, Mariss Karbon, Pat Maszk, Melanie Smith, Cherie O'Boyle, and K. Suh. "The Relations of Emotionality and Regulation to Dispositional and Situational Empathy-Related Responding." *Journal of Personality & Social Psychology* 66, no. 4 (1994): 776–97.

Eisenberg, Nancy, and Randy Lennon. "Sex Differences in Empathy and Related Capacities." *Psychological Bulletin* 94, no. 1 (1983): 100–131.

Eisenberg, Nancy, and Janet Strayer, eds. *Empathy and Its Development*. New York: Cambridge University Press, 1987.

Erwin, Douglas H., and Eric H. Davidson. "The Evolution of Hierarchical Gene Regulatory Networks." *Nature Reviews Genetics* 10, no. 2 (February 2009): 141–48. https://doi.org/10.1038/nrg2499.

Farwell, Lisa, and Bernard Weiner. "Bleeding Hearts and the Heartless: Popular Perceptions of Liberal and Conservative Ideologies." *Personality and Social Psychology Bulletin* 26, no. 7 (September 2000): 845–52. https://doi.org/10.1177/0146167200269009.

Fehr, Ernst, and Colin F. Camerer. "Social Neuroeconomics: The Neural Circuitry of Social Preferences." *Trends in Cognitive Sciences* 11, no. 10 (October 2007): 419–27.

Fehr, Ernst, and Urs Fischbacher. "The Nature of Human Altruism." *Nature* 425, no. 6960 (October 23, 2003): 785–91.

Fehr, Ernst, Urs Fischbacher, and Simon Gächter. "Strong Reciprocity, Human Cooperation, and the Enforcement of Social Norms." *Human Nature* 13, no. 1 (2002): 1–25. https://doi.org/10.1007/s12110-002-1012-7.

Fehr, Ernst, and Simon Gächter. "Altruistic Punishment in Humans." *Nature* 415, no. 6868 (January 10, 2002): 137–40.

Fischer, Peter, Tobias Greitemeyer, Fabian Pollozek, and Dieter Frey. "The Unresponsive Bystander: Are Bystanders More Responsive in Dangerous Emergencies?" *European Journal of Social Psychology* 36, no. 2 (March 2006): 267–78. https://doi.org/10.1002/ejsp.297.

Fischer, Peter, Joachim I. Krueger, Tobias Greitemeyer, Claudia Vogrincic, Andreas Kastenmüller, Dieter Frey, Moritz Heene, Magdalena Wicher, and Martina Kainbacher. "The Bystander-Effect: A Meta-Analytic Review on Bystander Intervention in Dangerous and Non-Dangerous Emergencies." *Psychological Bulletin* 137, no. 4 (2011): 517–37. https://doi.org/10.1037/a0023304.

Fleming, Alison S., Carl Corter, Joy Stallings, and Meir Steiner. "Testosterone and Prolactin Are Associated with Emotional Responses to Infant Cries in New Fathers." *Hormones and Behavior* 42, no. 4 (2002): 399–413.

Fleming, Alison S., Michael Numan, and Robert S. Bridges. "Father of Mothering: Jay S. Rosenblatt." *Hormones and Behavior* 55, no. 4 (April 2009): 484–87. https://doi.org/10.1016/j.yhbeh.2009.01.001.

Fleming, Alison S., and Jay S. Rosenblatt. "Olfactory Regulation of Maternal Behavior in Rats: II. Effects of Peripherally Induced Anosmia and Lesions of the Lateral Olfactory Tract in Pup-Induced Virgins." *Journal of Comparative and Physiological Psychology* 86 (1974): 233–46.

Fletcher, Harvey, and W. A. Munson. "Loudness, Its Definition, Measurement and Calculation." *Journal of the Acoustical Society of America* 5 (1933): 82–108.

Fraser, Orlaith N., and Thomas Bugnyar. "Do Ravens Show Consolation? Responses to Distressed Others." *PLoS ONE* 5, no. 5 (May 12, 2010): e10605. https://doi.org/10.1371/journal.pone.0010605.

Frederick, Shane, George Loewenstein, and Ted O'Donoghue. "Time Discounting and Time Preference: A Critical Review." *Journal of Economic Literature* 40, no. 2 (June 2002): 351–401. https://doi.org/10.1257/jel.40.2 .351.

Gansberg, Martin. "Thirty-Eight Who Saw Murder Didn't Call the Police." *New York Times*, March 27, 1964.

Gintis, Herbert. "Strong Reciprocity and Human Sociality." *Journal of Theoretical Biology* 206, no. 2 (2000): 169–79. https://doi.org/10.1006/jtbi.2000 .2111.

Gold, Joshua I., and Michael N. Shadlen. "Banburismus and the Brain." *Neuron* 36, no. 2 (October 2002): 299–308. https://doi.org/10.1016/S0896 -6273(02)00971-6.

Golle, Jessika, Stephanie Lisibach, Fred W. Mast, and Janek S. Lobmaier. "Sweet Puppies and Cute Babies: Perceptual Adaptation to Babyfacedness Transfers Across Species." *PLoS ONE* 8, no. 3 (March 13, 2013): e58248. https://doi.org/10.1371/journal.pone.0058248.

Goodson, James L., Aubrey M. Kelly, and Marcy A. Kingsbury. "Evolving Nonapeptide Mechanisms of Gregariousness and Social Diversity in Birds." *Hormones and Behavior* 61, no. 3 (March 2012): 239–50. https://doi .org/10.1016/j.yhbeh.2012.01.005.

Gould, James L. *Ethology: The Mechanisms and Evolution of Behavior.* New York: Norton, 1982.

Gray, Kurt, Adrian F. Ward, and Michael I. Norton. "Paying It Forward: Generalized Reciprocity and the Limits of Generosity." *Journal of Experimental Psychology: General* 143, no. 1 (2014): 247–54. https://doi.org/10.1037 /a0031047.

Grinstead, Charles M., and J. Laurie Snell. "Chapter 9: Central Limit Theorem." In *Introduction to Probability,* 2nd ed. Providence, RI: American Mathematical Society, 1997.

Gross, James J., and Robert W. Levenson. "Emotion Elicitation Using Films." *Cognition & Emotion* 9, no. 1 (January 1995): 87–108. https://doi.org/10 .1080/02699939508408966.

Gupta, Sandeep, Reshma Maurya, Monika Saxena, and Jonaki Sen. "Defining Structural Homology between the Mammalian and Avian Hippocampus through Conserved Gene Expression Patterns Observed in the Chick Embryo." *Developmental Biology* 366, no. 2 (June 15, 2012): 125–41. https://doi.org/10.1016/j.ydbio.2012.03.027.

Gustafson, Gwen E., and James A. Green. "On the Importance of Fundamental Frequency and Other Acoustic Features in Cry Perception and Infant Development." *Child Development* 60, no. 4 (1989): 772–80.

Hamilton, William D. "The Evolution of Altruistic Behavior." *The American Naturalist* 97, no. 896 (1963): 354–56.

——. "The Genetical Evolution of Social Behavior II." *Journal of Theoretical Biology* 7 (1964): 1–52.

Hansen, Stefan. "Maternal Behavior of Female Rats with 6-OHDA Lesions in the Ventral Striatum: Characterization of the Pup Retrieval Deficit." *Physiology & Behavior* 55, no. 4 (April 1994): 615–20. https://doi .org/10.1016/0031-9384(94)90034-5.

Harbaugh, William T., Ulrich Mayr, and Daniel R. Burghart. "Neural Responses to Taxation and Voluntary Giving Rebel Motives for Charitable Donation." *Science* 316 (2007): 1622–25.

Hauser, David J., Stephanie D. Preston, and R. Brent Stansfield. "Altruism in the Wild: When Affiliative Motives to Help Positive People Overtake Empathic Motives to Help the Distressed." *Journal of Experimental Psychology: General* 143, no. 3 (December 23, 2014): 1295–1305. https://doi .org/10.1037/a0035464.

Healy, Susan, Selvino R. Dekort, and Nicola S. Clayton. "The Hippocampus, Spatial Memory and Food Hoarding: A Puzzle Revisited." *Trends
in Ecology & Evolution* 20, no. 1 (January 2005): 17–22. https://doi.org/10
.1016/j.tree.2004.10.006.

Heise, Lori L. "Violence Against Women: An Integrated, Ecological Framework." *Violence Against Women* 4, no. 3 (June 1998): 262–90. https://doi
.org/10.1177/1077801298004003002.

Hepper, Peter G. "Kin Recognition: Functions and Mechanisms, a Review."
Biological Reviews 61, no. 1 (February 1986): 63–93. https://doi.org/10.1111
/j.1469-185X.1986.tb00427.x.

Hershfield, Hal E., Taya R. Cohen, and Leigh Thompson. "Short Horizons
and Tempting Situations: Lack of Continuity to Our Future Selves Leads
to Unethical Decision Making and Behavior." *Organizational Behavior
and Human Decision Processes* 117, no. 2 (March 2012): 298–310. https://
doi.org/10.1016/j.obhdp.2011.11.002.

Hertzberg, Hendrik. "Second Those Emotions: Hillary's Tears." *The New
Yorker*, January 21, 2008.

Hofer, Myron A. "Multiple Regulators of Ultrasonic Vocalization in the
Infant Rat." *Psychoneuroendocrinology* 21, no. 2 (February 1996): 203–17.
https://doi.org/10.1016/0306-4530(95)00042-9.

Hoffman, Kelly M., Sophie Trawalter, Jordan R. Axt, and M. Norman Oli-
ver. "Racial Bias in Pain Assessment and Treatment Recommendations,
and False Beliefs About Biological Differences Between Blacks and
Whites." *Proceedings of the National Academy of Sciences of the United States
of America* 113, no. 16 (April 19, 2016): 4296–301. https://doi.org/10.1073
/pnas.1516047113.

Hoffman, Martin L. "Empathy: Its Development and Prosocial Implications." *Nebraska Symposium on Motivation* 25 (1977): 169–217.

——. *Empathy and Moral Development: Implications for Caring and Justice.*
New York: Cambridge University Press, 2000.

——. "Is Altruism Part of Human Nature?" *Journal of Personality and Social
Psychology* 40 (1981): 121–37.

Holliday, Ruth, and Joanna Elfving-Hwang. "Gender, Globalization and
Aesthetic Surgery in South Korea." *Body & Society* 18, no. 2 (June 2012):
58–81. https://doi.org/10.1177/1357034X12440828.

Hrdy, Sarah Blaffer. *Mothers and Others*. Cambridge, MA: Harvard University Press, 2009.

Hume, David. *A Treatise of Human Nature*. North Chelmsford, MA: Courier Corporation, 2003.

Ickes, William John, ed. *Empathic Accuracy*. New York: Guilford Press, 1997.

Insel, Thomas R., and Larry J. Young. "The Neurobiology of Attachment." *Nature Reviews Neuroscience* 2, no. 2 (February 2001): 129–36.

Insel, Thomas R., and Carroll R. Harbaugh. "Lesions of the Hypothalamic Paraventricular Nucleus Disrupt the Initiation of Maternal Behavior." *Physiology & Behavior* 45 (1989): 1033–41.

Insel, Thomas R., Stephanie D. Preston, and James T. Winslow. "Mating in the Monogamous Male: Behavioral Consequences." *Physiology & Behavior* 57, no. 4 (1995): 615–27.

Israel, Salomon, Elad Lerer, Idan Shalev, Florina Uzefovsky, Mathias Reibold, Rachel Bachner-Melman, Roni Granot, et al. "Molecular Genetic Studies of the Arginine Vasopressin 1a Receptor (AVPR1a) and the Oxytocin Receptor (OXTR) in Human Behaviour: From Autism to Altruism with Some Notes in Between." *Progress in Brain Research* 170 (2008): 435–49.

Itard, Jean Marc Gaspard, and François Dagognet. *Victor de l'Aveyron*. Paris: Editions Allia, 1994.

Jameson, Tina L., John M. Hinson, and Paul Whitney. "Components of Working Memory and Somatic Markers in Decision Making." *Psychonomic Bulletin & Review* 11, no. 3 (2004): 515–20.

Jenni, Karen, and George Loewenstein. "Explaining the Identifiable Victim Effect." *Journal of Risk and Uncertainty* 14, no. 3 (1997): 235–57. https://doi.org/10.1023/A:1007740225484.

Johansson, Peter, and Margaret Kerr. "Psychopathy and Intelligence: A Second Look." *Journal of Personality Disorders* 19, no. 4 (August 2005): 357–69. https://doi.org/10.1521/pedi.2005.19.4.357.

Johnson, Ronald C. "Attributes of Carnegie Medalists Performing Acts of Heroism and of the Recipients of These Acts." *Ethology and Sociobiology* 17, no. 5 (September 1996): 355–62.

Johnstone, Rufus A. "Sexual Selection, Honest Advertisement and the Handicap Principle: Reviewing the Evidence." *Biological Reviews* 70 (1995): 1–65.

Jones, Alice. "The Psychology of Sustainability: What Planners Can Learn from Attitude Research." *Journal of Planning Education and Research* 16, no. 1 (September 1996): 56–65. https://doi.org/10.1177/0739456X9601600107.

Kabelik, David, and D. Sumner Magruder. "Involvement of Different Mesotocin (Oxytocin Homologue) Populations in Sexual and Aggressive
Behaviours of the Brown Anole." *Biology Letters* 10, no. 8 (August 31, 2014):
20140566. https://doi.org/10.1098/rsbl.2014.0566.

Kahneman, Daniel. *Thinking, Fast and Slow.* New York: Farrar, Straus and
Giroux, 2011.

Kaul, Padma, Paul W. Armstrong, Sunil Sookram, Becky K. Leung, Neil
Brass, and Robert C. Welsh. "Temporal Trends in Patient and Treatment
Delay Among Men and Women Presenting with ST-Elevation Myocardial Infarction." *American Heart Journal* 161, no. 1 (January 2011): 91–97.
https://doi.org/10.1016/j.ahj.2010.09.016.

Keating, Caroline F. "Do Babyfaced Adults Receive More Help? The (Cross-
Cultural) Case of the Lost Resume." *Journal of Nonverbal Behavior* 27,
no. 2 (2003): 89–109. https://doi.org/10.1023/A:1023962425692.

Keczer, Zsolt, Bálint File, Gábor Orosz, and Philip G. Zimbardo. "Social
Representations of Hero and Everyday Hero: A Network Study from Representative Samples." *PLOS ONE* 11, no. 8 (August 15, 2016): e0159354.
https://doi.org/10.1371/journal.pone.0159354.

Keller, Heidi, and Hiltrud Otto. "The Cultural Socialization of Emotion
Regulation During Infancy." *Journal of Cross-Cultural Psychology* 40,
no. 6 (November 2009): 996–1011. https://doi.org/10.1177/00220221093
48576.

Kendrick, Keith M., Ana PC Da Costa, Kevin D. Broad, Satoshi Ohkura,
Rosalinda Guevara, Frederic Lévy, and E. Barry Keverne. "Neural Con-
trol of Maternal Behaviour and Olfactory Recognition of Offspring."
Brain Research Bulletin 44, no. 4 (1997): 383–95.

Kilner, Rebecca M., David G. Noble, and Nicholas B. Davies. "Signals of
Need in Parent-Offspring Communication and Their Exploitation by the
Common Cuckoo." *Nature* 397, no. 6721 (1999): 667–72.

Knill, David C., and Alexandre Pouget. "The Bayesian Brain: The Role of
Uncertainty in Neural Coding and Computation." *Trends in Neurosciences*
27, no. 12 (December 2004): 712–19. https://doi.org/10.1016/j.tins.2004.10
.007.

Knoch, Daria, Michael A. Nitsche, Urs Fischbacher, Christoph Eisenegger,
Alvaro Pascual-Leone, and Ernst Fehr. "Studying the Neurobiology of Social
Interaction with Transcranial Direct Current Stimulation—The Example of
Punishing Unfairness." *Cerebral Cortex* 18, no. 9 (September 2008): 1987–90.

Knoch, D., A. Pascual-Leone, K. Meyer, V. Treyer, and E. Fehr. "Diminishing Reciprocal Fairness by Disrupting the Right Prefrontal Cortex." *Science* 314, no. 5800 (November 3, 2006): 829–32.

Koenigs, Michael, and Daniel Tranel. "Irrational Economic Decision-Making After Ventromedial Prefrontal Damage: Evidence from the Ultimatum Game." *Journal of Neuroscience* 27, no. 4 (January 24, 2007): 951–56.

Kogut, Tehila, and Ilana Ritov. "The 'Identified Victim' Effect: An Identified Group, or Just a Single Individual?" *Journal of Behavioral Decision Making* 18, no. 3 (July 2005): 157–67. https://doi.org/10.1002/bdm.492.

Kosfeld, Michael, Markus Heinrichs, Paul J. Zak, Urs Fischbacher, and Ernst Fehr. "Oxytocin Increases Trust in Humans." *Nature* 435, no. 7042 (June 2, 2005): 673–76.

Krain, Amy L., Amanda M. Wilson, Robert Arbuckle, F. Xavier Castellanos, and Michael P. Milham. "Distinct Neural Mechanisms of Risk and Ambiguity: A Meta-Analysis of Decision-Making." *NeuroImage* 32, no. 1 (2006): 477–84.

Krebs, John R. "Food-Storing Birds: Adaptive Specialization in Brain and Behaviour?" *Philosophical Transactions of the Royal Society of London. Series B: Biological Sciences* 329, no. 1253 (August 29, 1990): 153–60. https://doi.org/10.1098/rstb.1990.0160.

Krueger, Frank, Kevin McCabe, Jorge Moll, Nikolaus Kriegeskorte, Roland Zahn, Maren Strenziok, Armin Heinecke, and Jordan Grafman. "Neural Correlates of Trust." *Proceedings of the National Academy of Sciences USA* 104, no. 50 (December 11, 2007): 20084–89.

Kuraguchi, Kana, Kosuke Taniguchi, and Hiroshi Ashida. "The Impact of Baby Schema on Perceived Attractiveness, Beauty, and Cuteness in Female Adults." *SpringerPlus* 4, no. 1 (December 2015): 164. https://doi.org/10.1186/s40064-015-0940-8.

Lamm, Claus, Jean Decety, and Tania Singer. "Meta-Analytic Evidence for Common and Distinct Neural Networks Associated with Directly Experienced Pain and Empathy for Pain." *Neuroimage* 54, no. 3 (2011): 2492–502.

Latané, Bibb, and John M. Darley. "Bystander 'Apathy.'" *American Scientist* 57, no. 2 (1969): 244–68.

Lea, Stephen E. G., and Paul Webley. "Money as Tool, Money as Drug: The Biological Psychology of a Strong Incentive." *Behavioral and Brain Sciences* 29, no. 2 (2006): 161–209.

Lerner, Jennifer S., Ye Li, Piercarlo Valdesolo, and Karim S. Kassam. "Emotion and Decision Making." *Annual Review of Psychology* 66, no. 1 (January 3, 2015): 799–823. https://doi.org/10.1146/annurev-psych-010213 -115043

Levine, Mark, and Simon Crowther. "The Responsive Bystander: How Social Group Membership and Group Size Can Encourage as Well as Inhibit Bystander Intervention." *Journal of Personality and Social Psychology* 95, no. 6 (2008): 1429–39. https://doi.org/10.1037/a0012634.

Levy, Frédéric, Matthieu Keller, and Pascal Poindron. "Olfactory Regulation of Maternal Behavior in Mammals." *Hormones and Behavior* 46, no. 3 (September 2004): 284–302.

Liakos, Matthew, and Puja B. Parikh. "Gender Disparities in Presentation, Management, and Outcomes of Acute Myocardial Infarction." *Current Cardiology Reports* 20, no. 8 (August 2018): 64. https://doi.org/10.1007 /s11886-018-1006-7.

Lichtenstein, Gabriela. "Selfish Begging by Screaming Cowbirds, a Mimetic Brood Parasite of the Bay-Winged Cowbird." *Animal Behaviour* 61, no. 6 (2001): 1151–58.

Ligon, J. David, and D. Brent Burt. "Evolutionary Origins." In *Ecology and Evolution of Cooperative Breeding in Birds*, ed. Walter D. Koenig and Janis L. Dickinson, 5–34. Cambridge: Cambridge University Press, 2004.

Loewenstein, George F., Elke U. Weber, Christopher K. Hsee, and Ned Welch. "Risk as Feelings." *Psychological Bulletin* 127, no. 2 (2001): 267–86. https://doi.org/10.1037/0033-2909.127.2.267.

Loken, Line S., Johan Wessberg, India Morrison, Francis McGlone, and Hakan Olausson. "Coding of Pleasant Touch by Unmyelinated Afferents in Humans." *Nature Neuroscience* 12, no. 5 (2009): 547–48.

Lonstein, Joseph S., and Alison S. Fleming. "Parental Behaviors in Rats and Mice." *Current Protocols in Neuroscience* 15 (2002): Unit 8.15.

Lonstein, Joseph S., and Joan I. Morrell. "Neuroendocrinology and Neurochemistry of Maternal Motivation and Behavior." In *Handbook of Neurochemistry and Molecular Neurobiology*, ed. Abel Lajtha and Jeffrey D. Blaustein, 3rd ed., 195–245. Berlin: Springer-Verlag, 2007. http://www .springerlink.com/content/nw8357tv143w4w21/.

Lorberbaum, Jeffrey P., John D. Newman, Judy R. Dubno, Amy R. Horwitz, Ziad Nahas, Charlotte C. Teneback, Courtnay W. Bloomer, et al.

"Feasibility of Using FMRI to Study Mothers Responding to Infant Cries." *Depression and Anxiety* 10, no. 3 (1999): 99–104.

Lorberbaum, Jeffrey P., John D. Newman, Amy R. Horwitz, Judy R. Dubno, R. Bruce Lydiard, Mark B. Hamner, Daryl E. Bohning, and Mark S. George. "A Potential Role for Thalamocingulate Circuitry in Human Maternal Behavior." *Biological Psychiatry* 51, no. 6 (2002): 431–45.

Lorenz, Konrad. "Die Angeborenen Formen Möglicher Erfahrung [The Innate Forms of Potential Experience]." *Zeitschrift für Tierpsychologie* 5 (1943): 233–519.

——. *Studies in Animal and Human Behaviour: II.* Cambridge, MA: Harvard University Press, 1971.

Lorenz, Konrad, and Nikolaas Tinbergen. "Taxis und Instinkthandlung in der Eirollbewegung der Graugans [Directed and Instinctive Behavior in the Egg Rolling Movements of the Gray Goose]." *Zeitschrift für Tierpsychologie* 2 (1938): 1–29.

Lundström, Johan N., Annegret Mathe, Benoist Schaal, Johannes Frasnelli, Katharina Nitzsche, Johannes Gerber, and Thomas Hummel. "Maternal Status Regulates Cortical Responses to the Body Odor of Newborns." *Frontiers in Psychology* 4 (September 5, 2013). https://doi.org/10.3389/fpsyg.2013.00597.

Lynn, Spencer K., Jolie B. Wormwood, Lisa F. Barrett, and Karen S. Quigley. "Decision Making from Economic and Signal Detection Perspectives: Development of an Integrated Framework." *Frontiers in Psychology*, July 8, 2015. https://doi.org/10.3389/fpsyg.2015.00952.

Macht, Michael, and Jochen Mueller. "Immediate Effects of Chocolate on Experimentally Induced Mood States." *Appetite* 49, no. 3 (November 2007): 667–74. https://doi.org/10.1016/j.appet.2007.05.004.

MacLean, Paul D. "Brain Evolution Relating to Family, Play, and the Separation Call." *Archives of General Psychiatry* 42, no. 4 (1985): 405–17.

——. "The Brain in Relation to Empathy and Medical Education." *Journal of Nervous and Mental Disease* 144, no. 5 (1967): 374–82. https://doi.org/10.1097/00005053-196705000-00005.

——. *The Triune Brain in Evolution: Role in Paleocerebral Functions.* New York: Plenum Press, 1990.

Maestripieri, Dario. "The Biology of Human Parenting: Insights from Nonhuman Primates." *Neuroscience & Biobehavioral Reviews* 23, no. 3 (1999): 411–22. https://doi.org/10.1016/S0149-7634(98)00042-6.

Maestripieri, Dario, and Julia L. Zehr. "Maternal Responsiveness Increases During Pregnancy and After Estrogen Treatment in Macaques." *Hormones and Behavior* 34, no. 3 (1998): 223–30. https://doi.org/10.1006/hbeh .1998.1470.

Mampe, Birgit, Angela D. Friederici, Anne Christophe, and Kathleen Wermke. "Newborns' Cry Melody Is Shaped by Their Native Language." *Current Biology* 19, no. 23 (December 2009): 1994–97. https://doi.org/10 .1016/j.cub.2009.09.064.

Manning, Rachel, Mark Levine, and Alan Collins. "The Kitty Genovese Murder and the Social Psychology of Helping: The Parable of the 38 Witnesses." *American Psychologist* 62, no. 6 (2007): 555.

Marikar, Sheila. "Natasha Richardson Died of Epidural Hematoma After Skiing Accident." ABC News, March 19, 2009.

Marino, Lori, James K. Rilling, Shinko K. Lin, and Sam H. Ridgway. "Relative Volume of the Cerebellum in Dolphins and Comparison with Anthropoid Primates." *Brain, Behavior and Evolution* 56, no. 4 (2000): 204–11. https://doi.org/10.1159/000047205.

Marsh, Abigail A. "Neural, Cognitive, and Evolutionary Foundations of Human Altruism." *Wiley Interdisciplinary Reviews: Cognitive Science* 7, no. 1 (2016): 59–71.

Marsh, Abigail A., and R. James R. Blair. "Deficits in Facial Affect Recognition Among Antisocial Populations: A Meta-Analysis." *Neuroscience and Biobehavioral Reviews* 32 (2008): 454–65.

Marsh, Abigail A., and Robert E. Kleck. "The Effects of Fear and Anger Facial Expressions on Approach- and Avoidance-Related Behaviors." *Emotion* 5, no. 1 (2005): 119–24.

Marsh, Abigail A., Megan N. Kozak, and Nalini Ambady. "Accurate Identification of Fear Facial Expressions Predicts Prosocial Behavior." *Emotion* 7, no. 2 (2007): 239–51.

Martin, Loren J., Georgia Hathaway, Kelsey Isbester, Sara Mirali, Erinn L. Acland, Nils Niederstrasser, Peter M. Slepian, et al. "Reducing Social Stress Elicits Emotional Contagion of Pain in Mouse and Human Strangers." *Current Biology* 25, no. 3 (February 2015): 326–32. https://doi.org/10 .1016/j.cub.2014.11.028.

Mattson, Brandi J., Sharon E. Williams, Jay S. Rosenblatt, and Joan I. Morrell. "Comparison of Two Positive Reinforcing Stimuli: Pups and Cocaine Throughout the Postpartum Period." *Behavioral Neuroscience* 115 (2001): 683–94.

———. "Preferences for Cocaine or Pup-Associated Chambers Differentiates Otherwise Behaviorally Identical Postpartum Maternal Rats." *Psychopharmacology* 167 (2003): 1–8.

Maynard Smith, J. "Group Selection and Kin Selection." *Nature* 201 (1964): 1145–47.

McCabe, Kevin, Daniel Houser, Lee Ryan, Vernon Smith, and Theodore Trouard. "A Functional Imaging Study of Cooperation in Two-Person Reciprocal Exchange." *Proceedings of the National Academy of Sciences USA* 98, no. 20 (September 25, 2001): 11832–35.

McCarthy, Margaret M., Lee-Ming Kow, and Donald Wells Pfaff. "Speculations Concerning the Physiological Significance of Central Oxytocin in Maternal Behavior." *Annals of the New York Academy of Sciences* 652 (June 1992): 70–82. https://doi.org/10.1111/j.1749-6632.1992.tb34347.x.

McDougall, William. *An Introduction to Social Psychology.* London: Methuen, 1908.

McGuire, Anne M. "Helping Behaviors in the Natural Environment: Dimensions and Correlates of Helping." *Personality and Social Psychology Bulletin* 20, no. 1 (February 1994): 45–56. https://doi.org/10.1177/0146167294201004.

Meyer, Robyn J., Andreas A. Theodorou, and Robert A. Berg. "Childhood Drowning." *Pediatrics in Review* 27, no. 5 (May 2006): 163–69. https://doi.org/10.1542/pir.27-5-163.

Meyza, Ksenia Z., Inbal Ben-Ami Bartal, Marie H. Monfils, Jules B. Panksepp, and Ewelina Knapska. "The Roots of Empathy: Through the Lens of Rodent Models." *Neuroscience & Biobehavioral Reviews* 76 (May 2017): 216–34. https://doi.org/10.1016/j.neubiorev.2016.10.028.

Michelsson, K., K. Christensson, H. Rothgänger, and J. Winberg. "Crying in Separated and Non-Separated Newborns: Sound Spectrographic Analysis." *Acta Paediatrica* 85, no. 4 (April 1996): 471–75. https://doi.org/10.1111/j.1651-2227.1996.tb14064.x.

Milgram, Stanley. *Obedience to Authority: An Experimental View.* New York: Harper & Row, 1974.

Molenberghs, Pascal. "The Neuroscience of In-Group Bias." *Neuroscience & Biobehavioral Reviews* 37, no. 8 (September 2013): 1530–36. https://doi.org/10.1016/j.neubiorev.2013.06.002.

Moll, Jorge, Frank Krueger, Roland Zahn, Matteo Pardini, Ricardo de Oliveira-Souza, and Jordan Grafman. "Human Fronto-Mesolimbic Networks Guide Decisions About Charitable Donation." *Proceedings of the National Academy of Sciences USA* 103, no. 42 (October 17, 2006): 15623–28.

Moltz, Howard. "Contemporary Instinct Theory and the Fixed Action Pattern." *Psychological Review* 72, no. 1 (1965): 27–47. https://doi.org/10.1037/h0020275.

Mooradian, Todd A., Mark Davis, and Kurt Matzler. "Dispositional Empathy and the Hierarchical Structure of Personality." *American Journal of Psychology* 124, no. 1 (2011): 99. https://doi.org/10.5406/amerjpsyc.124.1.0099.

Morhenn, Vera B., Jang Woo Park, Elisabeth Piper, and Paul J. Zak. "Monetary Sacrifice Among Strangers Is Mediated by Endogenous Oxytocin Release after Physical Contact." *Evolution and Human Behavior* 29, no. 6 (2008): 375–83.

Morrison, I., Donna Lloyd, Giuseppe di Pellegrino, and Neil Roberts. "Vicarious Responses to Pain in Anterior Cingulate Cortex: Is Empathy a Multisensory Issue?" *Cognitive, Affective, and Behavioral Neuroscience* 4, no. 2 (June 2004): 270–78.

Mullins-Sweatt, Stephanie N., Natalie G. Glover, Karen J. Derefinko, Joshua D. Miller, and Thomas A. Widiger. "The Search for the Successful Psychopath." *Journal of Research in Personality* 44, no. 4 (August 2010): 554–58. https://doi.org/10.1016/j.jrp.2010.05.010.

Nave, Gideon, Colin Camerer, and Michael McCullough. "Does Oxytocin Increase Trust in Humans? A Critical Review of Research." *Perspectives on Psychological Science* 10, no. 6 (November 2015): 772–89. https://doi.org/10.1177/1745691615600138.

Newell, Ben R., and David R. Shanks. "Unconscious Influences on Decision Making: A Critical Review." *Behavioral and Brain Sciences* 37, no. 1 (February 2014): 1–19. https://doi.org/10.1017/S0140525X12003214.

Newsom, Jason T., and Richard Schulz. "Caregiving from the Recipient's Perspective: Negative Reactions to Being Helped." *Health Psychology* 17, no. 2 (1998): 172–81. https://doi.org/10.1037/0278-6133.17.2.172.

Nisbett, Richard E., and Timothy D. Wilson. "Telling More Than We Can Know: Verbal Reports on Mental Processes." *Psychological Review* 7 (1977): 231–59.

Nittono, Hiroshi, Michiko Fukushima, Akihiro Yano, and Hiroki Moriya. "The Power of Kawaii: Viewing Cute Images Promotes a Careful Behavior and Narrows Attentional Focus." *PLoS ONE* 7, no. 9 (September 26, 2012): e46362. https://doi.org/10.1371/journal.pone.0046362.

Nowak, Raymond, Matthieu Keller, David Val-Laillet, and Frédéric Lévy. "Perinatal Visceral Events and Brain Mechanisms Involved in the Development of Mother-Young Bonding in Sheep." *Hormones and Behavior* 52, no. 1 (2007): 92–98.

Nowbahari, Elise, and Karen L. Hollis. "Rescue Behavior: Distinguishing Between Rescue, Cooperation and Other Forms of Altruistic Behavior." *Communicative & Integrative Biology* 3, no. 2 (2010): 77–79.

Nowbahari, Elise, Alexandra Scohier, Jean-Luc Durand, and Karen L. Hollis. "Ants, Cataglyphis Cursor, Use Precisely Directed Rescue Behavior to Free Entrapped Relatives." *PLoS ONE* 4, no. 8 (August 12, 2009): e6573. https://doi.org/10.1371/journal.pone.0006573.

Numan, Michael. "Hypothalamic Neural Circuits Regulating Maternal Responsiveness Toward Infants." *Behavioral & Cognitive Neuroscience Reviews* 5, no. 4 (December 2006): 163–90.

——. "Motivational Systems and the Neural Circuitry of Maternal Behavior in the Rat." *Developmental Psychobiology* 49, no. 1 (January 2007): 12–21.

——. "Neural Circuits Regulating Maternal Behavior: Implications for Understanding the Neural Basis of Social Cooperation and Competition." In *Moving Beyond Self-Interest: Perspectives from Evolutionary Biology, Neuroscience, and the Social Sciences*, ed. Stephanie L. Brown, R. Michael Brown, and Louis A. Penner, 89–108. New York: Oxford University Press, 2011.

Numan, Michael, and Thomas R. Insel. *The Neurobiology of Parental Behavior*. New York: Springer, 2003.

Numan, Michael, Marilyn J. Numan, Jaclyn M. Schwarz, Christina M. Neuner, Thomas F. Flood, and Carl D. Smith. "Medial Preoptic Area Interactions with the Nucleus Accumbens-Ventral Pallidum Circuit and Maternal Behavior in Rats." *Behavioural Brain Research* 158, no. 1 (March 7, 2005): 53–68.

O'Connell, Sanjida M. "Empathy in Chimpanzees: Evidence for Theory of Mind?" *Primates* 36, no. 3 (1995): 397–410.

Oda, Ryo, Wataru Machii, Shinpei Takagi, Yuta Kato, Mia Takeda, Toko Kiyonari, Yasuyuki Fukukawa, and Kai Hiraishi. "Personality and Altruism in Daily Life." *Personality and Individual Differences* 56 (January 2014): 206–9. https://doi.org/10.1016/j.paid.2013.09.017.

O'Doherty, John. "Can't Learn Without You: Predictive Value Coding in Orbitofrontal Cortex Requires the Basolateral Amygdala." *Neuron* 39, no. 5 (August 28, 2003): 731–33.

Oliner, Samuel P. "Extraordinary Acts of Ordinary People." In *Altruism and Altruistic Love: Science, Philosophy, and Religion in Dialogue*, ed. Steven Post, Lynn G. Underwood, Jeffrey P. Schloss, and William B. Hurlburt, 123–39. Oxford: Oxford University Press, 2002.

Olkowicz, Seweryn, Martin Kocourek, Radek K. Lučan, Michal Porteš, W. Tecumseh Fitch, Suzana Herculano-Houzel, and Pavel Němec. "Birds Have Primate-like Numbers of Neurons in the Forebrain." *Proceedings of the National Academy of Sciences* 113, no. 26 (June 28, 2016): 7255–60. https://doi.org/10.1073/pnas.1517131113.

Öngür, Dost, and Joseph L. Price. "The Organization of Networks Within the Orbital and Medial Prefrontal Cortex of Rats, Monkeys and Humans." *Cerebral Cortex* 10 (2000): 206–19.

Oyserman, Daphna, Heather M. Coon, and Markus Kemmelmeier. "Rethinking Individualism and Collectivism: Evaluation of Theoretical Assumptions and Meta-Analyses." *Psychological Bulletin* 128, no. 1 (2002): 3–72. https://doi.org/10.1037/0033-2909.128.1.3.

Panksepp, Jules B., and Garet P. Lahvis. "Rodent Empathy and Affective Neuroscience." *Neuroscience & Biobehavioral Reviews* 35, no. 9 (October 2011): 1864–75. https://doi.org/10.1016/j.neubiorev.2011.05.013.

Peciña, Susana, and Kent C. Berridge. "Hedonic Hot Spot in Nucleus Accumbens Shell: Where Do μ-Opioids Cause Increased Hedonic Impact of Sweetness?" *Journal of Neuroscience* 25, no. 50 (2005): 11777–86.

Pedersen, Cort A., Jack D. Caldwell, Gary Peterson, Cheryl H. Walker, and George A. Mason. "Oxytocin Activation of Maternal Behavior in the Rata." *Annals of the New York Academy of Sciences* 652, no. 1 (2006): 58–69.

Pedersen, Cort A., Jack D. Caldwell, Cheryl Walker, Gail Ayers, and George A. Mason. "Oxytocin Activates the Postpartum Onset of Rat Maternal Behavior in the Ventral Tegmental and Medial Preoptic Areas." *Behavioral Neuroscience* 108 (1994): 1163–71.

Peng, Kaiping, and Richard E. Nisbett. "Culture, Dialectics, and Reasoning About Contradiction." *American Psychologist* 54, no. 9 (1999): 741–54. https://doi.org/10.1037/0003-066X.54.9.741.

Penner, Louis A., Barbara A. Fritzsche, J. Philip Craiger, and Tamara R. Freifeld. "Measuring the Prosocial Personality." *Advances in Personality Assessment* 10 (1995): 147–63.

Pepperberg, Irene. *The Alex Studies: Cognitive and Communicative Abilities of Grey Parrots.* Cambridge, MA: Harvard University Press, 2009.

Perez, Emilie C., Julie E. Elie, Ingrid C. A. Boucaud, Thomas Crouchet, Christophe O. Soulage, Hédi A. Soula, Frédéric E. Theunissen, and Clémentine Vignal. "Physiological Resonance Between Mates Through Calls as Possible Evidence of Empathic Processes in Songbirds." *Hormones and Behavior* 75 (September 1, 2015): 130–41. https://doi.org/10.1016/j.yhbeh.2015.09.002.

Pillemer, Karl, and David W. Moore. "Abuse of Patients in Nursing Homes: Findings from a Survey of Staff." *The Gerontologist* 29, no. 3 (June 1, 1989): 314–20. https://doi.org/10.1093/geront/29.3.314.

Pistole, Carole M. "Adult Attachment Styles: Some Thoughts on Closeness-Distance Struggles." *Family Process* 33, no. 2 (1994): 147–59. https://doi.org/10.1111/j.1545-5300.1994.00147.x.

Pletcher, Mark J., Stefan G. Kertesz, Michael A. Kohn, and Ralph Gonzales. "Trends in Opioid Prescribing by Race/Ethnicity for Patients Seeking Care in US Emergency Departments." *Journal of the American Medical Association* 299, no. 1 (January 2, 2008): 70–78. https://doi.org/10.1001/jama.2007.64.

Potegal, Michael, and John F. Knutson. *The Dynamics of Aggression: Biological and Social Processes in Dyads and Groups*. Hillsdale, NJ: Erlbaum, 1994.

"Pregnant Woman Rescues Husband from Shark Attack in Florida." *BBC News*, September 24, 2020. https://www.bbc.com/news/world-us-canada-54280694.

Prescott, Sara L., Rajini Srinivasan, Maria Carolina Marchetto, Irina Grishina, Iñigo Narvaiza, Licia Selleri, Fred H. Gage, Tomek Swigut, and Joanna Wysocka. "Enhancer Divergence and Cis-Regulatory Evolution in the Human and Chimp Neural Crest." *Cell* 163, no. 1 (September 2015): 68–83. https://doi.org/10.1016/j.cell.2015.08.036.

Preston, Stephanie D. "The Evolution and Neurobiology of Heroism." In *The Handbook of Heroism and Heroic Leadership*, ed. S. T. Allison, G. R. Goethals, and R. M. Kramer. New York: Taylor & Francis/Routledge, 2016.

——. "The Origins of Altruism in Offspring Care." *Psychological Bulletin* 139, no. 6 (2013): 1305–41. https://doi.org/10.1037/a0031755.

——. "The Rewarding Nature of Social Contact." *Science (New York, N.Y.)* 357, no. 6358 (29 2017): 1353–54. https://doi.org/10.1126/science.aao7192.

Preston, Stephanie D., Antoine Bechara, Hanna Damasio, Thomas J. Grabowski, R. Brent Stansfield, Sonya Mehta, and Antonio R. Damasio. "The Neural Substrates of Cognitive Empathy." *Social Neuroscience* 2, nos. 3–4 (2007): 254–75. https://doi.org/10.1080/17470910701376902.

Preston, Stephanie D., and F. B. M. de Waal. "Empathy: Its Ultimate and Proximate Bases." *Behavioral and Brain Sciences* 25, no. 1 (2002): 1–71. https://doi.org/10.1017/S0140525X02000018.

Preston, Stephanie D., and Frans B. M. de Waal. "Altruism." In *The Handbook of Social Neuroscience*, ed. Jean Decety and John T. Cacioppo, 565–85. New York: Oxford University Press, 2011.

Preston, Stephanie D., Melanie Ermler, Logan A. Bickel, and Yuxin Lei. "Understanding Empathy and Its Disorders Through a Focus on the Neural Mechanism." *Cortex* 127 (2020): 347–70. https://doi.org/10.1016/j.cortex.2020.03.001.

Preston, Stephanie D., Alicia J. Hofelich, and R. Brent Stansfield. "The Ethology of Empathy: A Taxonomy of Real-World Targets of Need and Their Effect on Observers." *Frontiers in Human Neuroscience* 7, no. 488 (2013): 1–13. https://doi.org/10.3389/fnhum.2013.00488.

Preston, Stephanie D., Morten Kringelbach, and Brian Knutson, eds. *The Interdisciplinary Science of Consumption*. Cambridge, MA: MIT Press, 2014.

Preston, Stephanie D., Julia D. Liao, Theodore P. Toombs, Rainer Romero-Canyas, Julia Speiser, and Colleen M. Seifert. "A Case Study of a Conservation Flagship Species: The Monarch Butterfly." *Biodiversity and Conservation* 30 (2021): 2057–77.

Preston, Stephanie D., Tingting Liu, and Nadia R. Danienta. "Neoteny: The Adaptive Attraction Toward 'Cuteness' Across Ages and Domains." Forthcoming.

Preston, Stephanie D., and Andrew D. MacMillan-Ladd. "Object Attachment and Decision-Making." *Current Opinion in Psychology* 39 (June 2021): 31–37. https://doi.org/10.1016/j.copsyc.2020.07.01.

Preston, Stephanie D., Brian D. Vickers, Reiner Romero-Cayas, and Colleen M. Seifert. "Leveraging Differences in How Liberals versus Conservatives Think About the Earth Improves Pro-Environmental Responses." Forthcoming.

Qui, Linda. "5 Irresistible National Geographic Cover Photos," n.d. https://www.nationalgeographic.com/news/2014/12/141206-magazine-covers-photography-national-geographic-afghan-girl/.

Quiatt, Duane. "Aunts and Mothers: Adaptive Implications of Allomaternal Behavior of Nonhuman Primates." *American Anthropologist* 81, no. 2 (June 1979): 310–19. https://doi.org/10.1525/aa.1979.81.2.02a00040.

Rajwani, Naheed. "Study: Rats Are Nice to One Another." *Chicago Tribune*, January 15, 2014.

Reynolds, John D., Nicholas B. Goodwin, and Robert P. Freckleton. "Evolutionary Transitions in Parental Care and Live Bearing in Vertebrates." *Philosophical Transactions of the Royal Society of London. Series B: Biological Sciences* 357, no. 1419 (March 29, 2002): 269–81. https://doi.org/10.1098/rstb.2001.0930.

Rilling, James K., David A. Gutman, Thorsten R. Zeh, Giuseppe Pagnoni, Gregory S. Berns, and Clinton D. Kilts. "A Neural Basis for Social Cooperation." *Neuron* 35, no. 2 (2002): 395–405.

Rilling, James K., and Jennifer S. Mascaro. "The Neurobiology of Fatherhood." *Current Opinion in Psychology* 15 (June 2017): 26–32. https://doi.org/10.1016/j.copsyc.2017.02.013.

Rilling, James K., Alan G. Sanfey, Jessica A. Aronson, Leigh E. Nystrom, and Jonathan D. Cohen. "Opposing BOLD Responses to Reciprocated and Unreciprocated Altruism in Putative Reward Pathways." *Neuroreport* 15, no. 16 (2004): 2539–43.

Rizzolatti, Giacomo, Luciano Fadiga, Vittorio Gallese, and Leonardo Fogassi. "Premotor Cortex and the Recognition of Motor Actions." *Cognitive Brain Research* 3, no. 2 (March 1996): 131–41. https://doi.org/10.1016/0926-6410(95)00038-0.

Robbins, Trevor W. "Homology in Behavioural Pharmacology: An Approach to Animal Models of Human Cognition." *Behavioural Pharmacology* 9, no. 7 (November 1998): 509–19. https://doi.org/10.1097/00008877-1998 11000-00005.

Rooney, Patrick M. "The Growth in Total Household Giving Is Camouflaging a Decline in Giving by Small and Medium Donors: What Can We Do About It?" *Nonprofit Quarterly*, August 27, 2019. https://nonprofitquarterly.org/total-household-growth-decline-small-medium-donors.

Rosch, Eleanor. "Principles of Categorization." In *Cognition and Categorization*, ed. Eleanor Rosch and Barbara B. Lloyd, 27–48. Hillsdale, NJ: Erlbaum, 1978.

Rosenblatt, Jay S. "Nonhormonal Basis of Maternal Behavior in the Rat." *Science* 156 (1967): 1512–14.

Rosenblatt, Jay S., and Kensey Ceus. "Estrogen Implants in the Medial Preoptic Area Stimulate Maternal Behavior in Male Rats." *Hormones and Behavior* 33 (1998): 23–30.

Ross, Heather E., Sara M. Freeman, Lauren L. Spiegel, Xianghui Ren, Ernest F. Terwilliger, and Larry J. Young. "Variation in Oxytocin Receptor Density in the Nucleus Accumbens Has Differential Effects on Affiliative Behaviors in Monogamous and Polygamous Voles." *Journal of Neuroscience* 29, no. 5 (February 4, 2009): 1312–18. https://doi.org/10.1523 /JNEUROSCI.5039-08.2009.

Saltzman, Wendy, and Toni E. Ziegler. "Functional Significance of Hormonal Changes in Mammalian Fathers." *Journal of Neuroendocrinology* 26, no. 10 (October 2014): 685–96. https://doi.org/10.1111/jne.12176.

Sanfey, Alan G., James K. Rilling, Jessica A. Aronson, Leigh E. Nystrom, and Jonathan D. Cohen. "The Neural Basis of Economic Decision-Making in the Ultimatum Game." *Science* 300, no. 5626 (June 13, 2003): 1755–58.

Sapolsky, Robert M. "The Influence of Social Hierarchy on Primate Health." *Science* 308, no. 5722 (April 29, 2005): 648–52. https://doi.org/10.1126/science .1106477.

——. "Stress, Glucocorticoids, and Damage to the Nervous System: The Current State of Confusion." *Stress* 1, no. 1 (2009): 1–19. https://doi.org /10.3109/10253899609001092.

Schiefenhövel, Wulf. *Geburtsverhalten und Reproduktive Strategien der Eipo: Ergebnisse Humanethologischer und Ethnomedizinischer Untersuchungen im Zentralen Bergland von Irian Jaya (West-Neuguinea), Indonesien* [Birth Behavior and Reproductive Strategies of the Eipo: Results of Human Ethology and Ethnomedical Researches in the Central Highlands of Irian Jaya (West New Guinea), Indonesia]. Berlin: D. Reimer, 1988.

Schneirla, Theodore C. "An Evolutionary and Developmental Theory of Biphasic Processes Underlying Approach and Withdrawal." *Nebraska Symposium on Motivation* 7 (1959): 1–42.

Schoenbaum, Geoffrey, Andrea A. Chiba, and Michela Gallagher. "Orbitofrontal Cortex and Basolateral Amygdala Encode Expected Outcomes During Learning." *Nature Neuroscience* 1, no. 2 (June 1998): 155–59.

Schultz, Wolfram. "Neural Coding of Basic Reward Terms of Animal Learning Theory, Game Theory, Microeconomics and Behavioural Ecology." *Current Opinion in Neurobiology* 14, no. 2 (April 2004): 139–47. https:// doi.org/10.1016/j.conb.2004.03.017.

Schulz, Horst, Gábor L. Kovács, and Gyula Telegdy. "Action of Posterior Pituitary Neuropeptides on the Nigrostriatal Dopaminergic System."

European Journal of Pharmacology 57, no. 2–3 (1979): 185–90. https://doi .org/10.1016/0014-2999(79)90364-9.

Schwarz, Norbert, and Gerald L. Clore. "Mood as Information: 20 Years Later." *Psychological Inquiry* 14, no. 3–4 (2003): 296–303.

Seyfarth, Robert, Dorothy L. Cheney, and Peter Marler. "Monkey Responses to Three Different Alarm Calls: Evidence of Predator Classification and Semantic Communication." *Science* 210, no. 4471 (November 14, 1980): 801–3. https://doi.org/10.1126/science.7433999.

Sherry, David F., M. R. Forbes, Moshe Khurgel, and Gwen O. Ivy. "Females Have a Larger Hippocampus Than Males in the Brood-Parasitic Brown-Headed Cowbird." *Proceedings of the National Academy of Sciences* 90, no. 16 (August 15, 1993): 7839–43. https://doi.org/10.1073/pnas.90.16.7839.

Sherry, David F., Lucia F. Jacobs, and Steven J. C. Gaulin. "Spatial Memory and Adaptive Specialization of the Hippocampus." *Trends in Neurosciences* 15, no. 8 (August 1992): 298–303. https://doi.org/10.1016/0166 -2236(92)90080-R.

Shiota, Michelle, Esther K. Papies, Stephanie D. Preston, and Disa A. Sauter. "Positive Affect and Behavior Change." *Current Opinion in Behavioral Sciences* 39 (2021): 222–28.

Siegel, Harold I., and Jay S. Rosenblatt. "Estrogen-Induced Maternal Behavior in Hysterectomized-Ovariectomized Virgin Rats." *Physiology & Behavior* 14, no. 4 (1975): 465–71.

Simonyan, Kristina, Barry Horwitz, and Erich D. Jarvis. "Dopamine Regulation of Human Speech and Bird Song: A Critical Review." *Brain and Language* 122, no. 3 (September 1, 2012): 142–50. https://doi.org/10.1016/j .bandl.2011.12.009.

Singer, Tania, Ben Seymour, John O'Doherty, Holger Kaube, Raymond J. Dolan, and Chris D. Frith. "Empathy for Pain Involves the Affective but Not Sensory Components of Pain." *Science* 303, no. 5661 (February 20, 2004): 1157–62.

Singer, Tania, Ben Seymour, John P. O'Doherty, Klaas E. Stephan, Raymond J. Dolan, and Chris D. Frith. "Empathic Neural Responses Are Modulated by the Perceived Fairness of Others." *Nature* 439, no. 7075 (January 26, 2006): 466–69.

Singer, Tania, Romana Snozzi, Geoffrey Bird, Predrag Petrovic, Giorgia Silani, Markus Heinrichs, and Raymond J. Dolan. "Effects of Oxytocin and Prosocial Behavior on Brain Responses to Direct and Vicariously Experienced Pain." *Emotion* 8, no. 6 (December 2008): 781–91.

Slotnick, Burton M. "Disturbances of Maternal Behavior in the Rat Following Lesions of the Cingulate Cortex." *Behaviour* 29, no. 2 (1967): 204–36.

Slovic, Paul. "If I Look at the Mass I Will Never Act: Psychic Numbing and Genocide." In *Emotions and Risky Technologies*, ed. Sabine Roeser, 5:37–59. Dordrecht: Springer Netherlands, 2010. https://doi.org/10.1007 /978-90-481-8647-1_3.

Slovic, Paul, and Ellen Peters. "Risk Perception and Affect." *Current Directions in Psychological Science* 15, no. 6 (December 2006): 322–25. https:// doi.org/10.1111/j.1467-8721.2006.00461.x.

Small, Deborah A., and George Loewenstein. "Helping a Victim or Helping the Victim: Altruism and Identifiability." *Journal of Risk and Uncertainty* 26, no. 1 (2003): 5–16. https://doi.org/10.1023/A:1022299422219.

Small, Deborah A., George Loewenstein, and Paul Slovic. "Sympathy and Callousness: The Impact of Deliberative Thought on Donations to Identifiable and Statistical Victims." *Organizational Behavior and Human Decision Processes* 102, no. 2 (March 2007): 143–53. https://doi.org/10.1016/j .obhdp.2006.01.005.

Smith, H. Lovell, Anthony Fabricatore, and Mark Peyrot. "Religiosity and Altruism Among African American Males: The Catholic Experience." *Journal of Black Studies* 29, no. 4 (March 1999): 579–97. https://doi.org/10 .1177/002193479902900407.

Smith, Sarah Francis, Scott O. Lilienfeld, Karly Coffey, and James M. Dabbs. "Are Psychopaths and Heroes Twigs off the Same Branch? Evidence from College, Community, and Presidential Samples." *Journal of Research in Personality* 47, no. 5 (October 2013): 634–46. https://doi.org/10.1016/j.jrp .2013.05.006.

Spear, Norman E., Winfred F. Hill, and Denis J. O'Sullivan. "Acquisition and Extinction After Initial Trials Without Reward." *Journal of Experimental Psychology* 69, no. 1 (1965): 25–29. https://doi.org/10.1037 /h0021628.

Stallings, Joy, Alison S. Fleming, Carl Corter, Carol Worthman, and Meir Steiner. "The Effects of Infant Cries and Odors on Sympathy, Cortisol, and Autonomic Responses in New Mothers and Nonpostpartum Women." *Parenting-Science and Practice* 1, nos. 1–2 (2001): 71–100.

Staub, Ervin. "A Child in Distress: The Influence of Nurturance and Modeling on Children's Attempts to Help." *Developmental Psychology* 5, no. 1 (1971): 124–32. https://doi.org/10.1037/h0031084.

Staub, Ervin, Daniel Bar-Tal, Jerzy Karylowski, and Janusz Reykowski, eds. *Development and Maintenance of Prosocial Behavior.* Boston: Springer, 1984. https://doi.org/10.1007/978-1-4613-2645-8.

Stern, Judith M., and Joseph S. Lonstein. "Neural Mediation of Nursing and Related Maternal Behaviors." *Progress in Brain Research* 133 (2001): 263–78.

Storey, Anne E., Carolyn J. Walsh, Roma L. Quinton, and Katherine E. Wynne-Edwards. "Hormonal Correlates of Paternal Responsiveness in New and Expectant Fathers." *Evolution and Human Behavior* 21 (2000): 79–95.

Strassmann, Joan E., Yong Zhu, and David C. Queller. "Altruism and Social Cheating in the Social Amoeba Dictyostelium Discoideum." *Nature* 408 (2000): 965–67.

Sullivan, Helen. "Florida Man Rescues Puppy from Jaws of Alligator Without Dropping Cigar." *The Guardian*, November 23, 2020. https://www.theguardian.com/us-news/2020/nov/23/man-rescues-puppy-from-alligator-without-dropping-cigar.

Swets, John A. *Signal Detection Theory and ROC Analysis in Psychology and Diagnostics: Collected Papers.* New York: Psychology Press, 2014.

Taylor, Katherine, Allison Visvader, Elise Nowbahari, and Karen L. Hollis. "Precision Rescue Behavior in North American Ants." *Evolutionary Psychology* 11, no. 3 (July 2013): 14747049301100. https://doi.org/10.1177/147470491301100312.

Taylor, Shelley E., Laura Cousino Klein, Brian P. Lewis, Tara L. Gruenewald, Regan A. R. Gurung, and John A. Updegraff. "Biobehavioral Responses to Stress in Females: Tend-and-Befriend, Not Fight-or-Flight." *Psychological Review* 107, no. 3 (2000): 411–29.

Tinbergen, Nikolaas. "On Aims and Methods of Ethology." *Zeitschrift für Tierpsychologie* 20 (1963): 410–33.

Tranel, Daniel, and Antonio R. Damasio. "The Covert Learning of Affective Valence Does Not Require Structures in Hippocampal System or Amygdala." *Journal of Cognitive Neuroscience* 5, no. 1 (January 1993): 79–88. https://doi.org/10.1162/jocn.1993.5.1.79.

Trivers, Robert L. "The Evolution of Reciprocal Altruism." *Quarterly Review of Biology* 46 (1971): 35–57.

Troisi, Alfonso, Filippo Aureli, Paola Piovesan, and Francesca R. D'Amato. "Severity of Early Separation and Later Abusive Mothering in Monkeys: What Is the Pathogenic Threshold?" *Journal of Child Psychology and Psychiatry* 30, no. 2 (March 1989): 277–84.

Van Anders, Sari M., Richard M. Tolman, and Brenda L. Volling. "Baby Cries and Nurturance Affect Testosterone in Men." *Hormones and Behavior* 61, no. 1 (2012): 31–36. https://doi.org/10.1016/j.yhbeh.2011.09.012.

Van IJzendoorn, Marinus H. "Attachment, Emergent Morality, and Aggression: Toward a Developmental Socioemotional Model of Antisocial Behaviour." *International Journal of Behavioral Development* 21, no. 4 (November 1997): 703–27. https://doi.org/10.1080/016502597384631.

Van IJzendoorn, Marinus H., and Marian J. Bakermans-Kranenburg. "A Sniff of Trust: Meta-Analysis of the Effects of Intranasal Oxytocin Administration on Face Recognition, Trust to In-Group, and Trust to Out-Group." *Psychoneuroendocrinology* 37, no. 3 (March 2012): 438–43. https://doi.org/10.1016/j.psyneuen.2011.07.008.

Vareikaite, Vaiva. "60 Times Florida Man Did Something So Crazy We Had to Read the Headings Twice." *Bored Panda*, 2018. https://www.boredpanda .com/hilarious-florida-man-headings/?utm_source=google&utm _medium=organic&utm_campaign=organic.

Verbeek, Peter, and Frans B. M. de Waal. "Peacemaking Among Preschool Children." *Peace and Conflict: Journal of Peace Psychology* 7, no. 1 (2001): 5–28. https://doi.org/10.1207/S15327949PAC0701_02.

Vickers, Brian D., Rachael D. Seidler, R. Brent Stansfield, Daniel H. Weissman, and Stephanie D. Preston. "Motor System Engagement in Charitable Giving: The Offspring Care Effect." Forthcoming.

Visalberghi, Elisabetta, and Elsa Addessi. "Seeing Group Members Eating a Familiar Food Enhances the Acceptance of Novel Foods in Capuchin Monkeys." *Animal Behaviour* 60, no. 1 (July 2000): 69–76. https://doi.org /10.1006/anbe.2000.1425.

Wagner, Allan R. "Effects of Amount and Percentage of Reinforcement and Number of Acquisition Trials on Conditioning and Extinction." *Journal of Experimental Psychology* 62, no. 3 (1961): 234–42. https://doi.org/10.1037 /h0042251.

Waldman, Bruce. "The Ecology of Kin Recognition." *Annual Review of Ecology and Systematics* 19, no. 1 (November 1988): 543–71. https://doi.org/10 .1146/annurev.es.19.110188.002551.

Warneken, Felix, and Michael Tomasello. "Varieties of Altruism in Children and Chimpanzees." *Trends in Cognitive Sciences* 13, no. 9 (2009): 397–402.

Warren, William H. "Perceiving Affordances: Visual Guidance of Stair Climbing." *Journal of Experimental Psychology: Human Perception and Performance* 10, no. 5 (1984): 683–703. https://doi.org/10.1037/0096-1523.10.5.683.

Wiesenfeld, Alan R., and Rafael Klorman. "The Mother's Psychophysiological Reactions to Contrasting Affective Expressions by Her Own and an Unfamiliar Infant." *Developmental Psychology* 14, no. 3 (1978): 294–304. https://doi.org/10.1037/0012-1649.14.3.294.

Wilkinson, Gerald S. "Food Sharing in Vampire Bats." *Scientific American* 262 (1990): 76–82.

Wilson, David S. "A Theory of Group Selection." *Proceedings of the National Academy of Sciences USA* 72, no. 1 (January 1975): 143–46.

Wilson, David S., and Lee A. Dugatkin. "Group Selection and Assortative Interactions." *The American Naturalist* 149, no. 2 (February 1, 1997): 336–51. https://doi.org/10.1086/285993.

Wilson, Edward O. "A Chemical Releaser of Alarm and Digging Behavior in the Ant Pogonomyrmex Badius (Latreille)." *Psyche* 65, no. 2–3 (1958): 41–51.

Wilson, Margo, Martin Daly, and Nicholas Pound. "An Evolutionary Psychological Perspective on the Modulation of Competitive Confrontation and Risk-Taking." *Hormones, Brain and Behavior* 5 (2002): 381–408.

Wilsoncroft, William E. "Babies by Bar-Press: Maternal Behavior in the Rat." *Behavior Research Methods, Instruments and Computers* 1 (1969): 229–30.

Wynne-Edwards, Katherine E. "Hormonal Changes in Mammalian Fathers." *Hormones and Behavior* 40, no. 2 (September 2001): 139–45. https://doi.org/10.1006/hbeh.2001.1699.

Wynne-Edwards, Katherine E., and Mary E. Timonin. "Paternal Care in Rodents: Weakening Support for Hormonal Regulation of the Transition to Behavioral Fatherhood in Rodent Animal Models of Biparental Care." *Hormones and Behavior* 52, no. 1 (2007): 114–21.

Zahn-Waxler, Carolyn, Barbara Hollenbeck, and Marian Radke-Yarrow. "The Origins of Empathy and Altruism." In *Advances in Animal Welfare*

Science, ed. M. W. Fox and L. D. Mickley, 21–39. Washington, DC: Humane Society of the United States, 1984.

Zahn-Waxler, Carolyn, and Marian Radke-Yarrow. "The Development of Altruism: Alternative Research Strategies." In *The Development of Prosocial Behavior*, ed. Nancy Eisenberg, 133–62. New York: Academic Press, 1982.

Zahn-Waxler, Carolyn, Marian Radke-Yarrow, and Robert A. King. "Child Rearing and Children's Prosocial Initiations Toward Victims of Distress." *Child Development* 50, no. 2 (1979): 319–30.

Zahn-Waxler, Carolyn, Marian Radke-Yarrow, Elizabeth Wagner, and Michael Chapman. "Development of Concern for Others." *Developmental Psychology* 28, no. 1 (1992): 126–36.

Zak, Paul J. "The Neurobiology of Trust." *Scientific American* 298, no. 6 (June 2008): 88–92, 95.

Zak, Paul J., Robert Kurzban, and William T. Matzner. "Oxytocin Is Associated with Human Trustworthiness." *Hormones and Behavior* 48, no. 5 (December 2005): 522–27.

Zak, Paul J., Angela A. Stanton, and Sheila Ahmadi. "Oxytocin Increases Generosity in Humans." *PLoS ONE* 2, no. 11 (2007): e1128.

Zaki, Jamil, Niall Bolger, and Kevin N. Ochsner. "It Takes Two: The Interpersonal Nature of Empathic Accuracy." *Psychological Science* 19, no. 4 (April 2008): 399–404. https://doi.org/10.1111/j.1467-9280.2008.02099.x.

Zebrowitz, Leslie A., Karen Olson, and Karen Hoffman. "Stability of Babyfaceness and Attractiveness Across the Life Span." *Journal of Personality and Social Psychology* 64, no. 3 (1993): 453–66. https://doi.org/10.1037/0022-3514.64.3.453.

Zeifman, Debra M. "An Ethological Analysis of Human Infant Crying: Answering Tinbergen's Four Questions." *Developmental Psychobiology* 39, no. 4 (2001): 265–85. https://doi.org/10.1002/dev.1005.

Ziegler, Toni E. "Hormones Associated with Non-Maternal Infant Care: A Review of Mammalian and Avian Studies." *Folia Primatologica* 71, no. 1–2 (2000): 6–21.

Ziegler, Toni E., and Charles T. Snowdon. "The Endocrinology of Family Relationships in Biparental Monkeys." In *The Endocrinology of Social Relationships*, ed. Peter T. Ellison and Peter B. Gray, 138–58. Cambridge, MA: Harvard University Press, 2009.

Zimbardo, Philip G. "On 'Obedience to Authority.'" *American Psychologist* 29, no. 7 (1974): 566–67. https://doi.org/10.1037/h0038158.

Zimbardo, Philip G., Christina Maslach, and Craig Haney. "Reflections on the Stanford Prison Experiment: Genesis, Transformations, Consequences." In *Obedience to Authority: Current Perspectives on the Milgram Paradigm*, ed. T. Blass, 193–237. Hoboken, NJ: Erlbaum, 1999.

Zsambok, Caroline E., and Gary A. Klein. *Naturalistic Decision Making*. Philadelphia: Erlbaum, 1997.

INDEX

abuse: child, 97–98, 166, 187; partner, 104

ACC. *See* anterior cingulate cortex (ACC)

active care, 16–17, 26, 28, 34, 36, 71, 73, 79, 113

acute need, 148

addiction, 30, 142

adoption, 59

advantageous choice, 213

affect: in decision making, 215; linkage, 76; in processing, 201; sharing, 16, 44, 198, 207, 211–12, 197. *See also* emotion

affiliation, 67

African gray parrot. *See* birds

against empathy, 94, 154. *See also* Bloom, Paul

aging, 143–44

agreeableness. *See* personality

aid rejection, 200, 142–43

alarm calls, 33, 57, 58, 67, 221; in ground squirrels, xii, 195, 203; in prairie dogs, 57, 67

alloparental care. *See* cooperative breeding

altricial development, 149, 227

altruistic personality. *See* personality

altruistic response, 171; and additive impact of attributes, 65, 20, 169; common concerns with the model, 11, 36, 91; error or mistake, 93, 221; functional properties, 32, 39, 63; threshold for responding, 189

altruistic urge. *See* urge

Alzheimer's disease, 143

amnesia, 121

amoeba, 90, 203, 219

amygdala, 15, 23–25, 42–43, 61, 109, 118–21, 124, 128, 130, 134–35, 187–88, 214

analogy, 38, 228

ancestral state, 230

anger, 212; in facial expression, 103; priming, 105

animal cruelty, 4

anterior cingulate cortex (ACC), 44, 61, 89, 124–25, 127, 132, 211

anterior hypothalamic nucleus (AHN), 24–25, 109, 125

anterior insula. *See* insula

anthropomorphism, 42

anticipation, 123, 126, 201, 214. *See also* prediction

Lightning Source UK Ltd.
Milton Keynes UK
UKHW040601170522
402980UK00007B/8/J